Report of the

CENTRE FOR OVERSEAS PEST RESEARCH

January-December 1975

Centre for Overseas Pest Research
College House, Wrights Lane
London W8 5SJ

©
Crown Copyright
1976

Short extracts from the text may be reproduced
provided that the source is acknowledged

ISBN 0 85135 085 2

Available only from the Centre for Overseas Pest Research

Price £2.75

Printed by Hobbs the Printers, Southampton
for Her Majesty's Stationery Office
(1296) Dd151327 1,500 12/76 G3313

CONTENTS

I GENERAL REVIEW

General review	1
Staff and accommodation	2
Outstanding events	2
Honours and awards	6

II INTERNATIONAL CO-OPERATION

A UNITED NATIONS AGENCIES

United Nations Development Programme	7
United Nations Environmental Programme	7
Food and Agriculture Organization	7
World Health Organization	9
World Meteorological Organization	10

B INTERNATIONAL ORGANIZATIONS

Association d'Acridologie	11
Association for the Advancement of Agricultural Sciences in Africa	11
Comité Européen de Normalisation	11
Desert Locust Control Organization for Eastern Africa	11
East African Agriculture and Forestry Research Organization	11
International Centre for Insect Physiology and Ecology	11
International Crops Research Institute for the Semi-Arid Tropics	12
International Red Locust Control Organization for Central and Southern Africa	12
International Rice Research Institute	13
Organisation Commune de lutte Antiacridienne et de Lutte Antiaviaire	13
Organization for Economic Cooperation and Development	13
Organisation Internationale Contre le Criquet Migrateur Africain	14

C INDIVIDUAL COUNTRIES

Australia	15
Bangladesh	15
Botswana	15
United Republic of Cameroun	15
Caribbean Islands	17
Central African Republic	17
Costa Rica	18
Egypt	18
Ethiopia	18
The Gambia	18

W. Germany	18
Ghana	20
Greece	20
Kenya	20
Malawi	21
Malaysia	21
Malta	21
Mauritius	22
Morocco	22
Nigeria	22
Pakistan	22
Sri Lanka	23
Sudan	23
Swaziland	23
Switzerland	23
Tanzania	25
Thailand	25
USA	25
Yemen Arab Republic	25

III COOPERATION WITH UK BASED ORGANIZATIONS

Animal Virus Research Institute	27
British Museum (Natural History)	27
British Standards Institution	27
Building Research Establishment, Princes Risborough Laboratory	27
Commonwealth Institute of Entomology	28
Commonwealth Institue of Helminthology	28
Commonwealth Mycological Institute	28
Glasshouse Crops Research Institute	28
Meteorological Office	28
Ministry of Agriculture, Fisheries and Food	29
Overseas Spraying Machinery Centre	29
Rothamsted Experimental Station	29
Royal Botanic Garden	29
Tropical Products Institute	30
Unit of Invertebrate Virology	30
Universities	30
Weed Research Organization	32

IV TECHNICAL REPORT

A LOCUSTS AND GRASSHOPPERS

1 Acridid taxonomy	33
2 Insect/host-plant interactions	34
3 Reproduction and development	36

4 Neurophysiology and flight	41
5 Desert Locust research	42
6 *Zonocerus variegatus* in Nigeria	43
7 Locusts and grasshoppers of economic importance in Thailand	44

B TERMITES

1 Biology and testing	46
2 Taxonomy and morphology	47
3 Identification and advisory services	47
4 Field research in Nigeria	48

C LEPIDOPTERA

1 Laboratory and field studies of *Spodoptera*	53
2 Biogeographical studies of the African Armyworm, *Spodoptera exempta*	59
3 *Chilo partellus* oviposition behaviour	59
4 Insect viruses	60

D METEOROLOGICAL AND RADAR STUDIES

1 Meteorology	62
2 Radar	64
3 Sudan Gezira Project	64

E CROP PROTECTION PROJECTS

1 Crop protection operations in the Sahel	66
2 Regional Plant Protection Team in the Caribbean	67
3 Multicropping systems research in Costa Rica	69

F PUBLIC HEALTH PROJECTS

1 Schistosomiasis	71
2 Control of river blindness in West Africa	75
3 Tsetse fly control	76

G VERTEBRATE PESTS

1 Birds	78
2 Rodents and small mammals	79

H PEST CONTROL CHEMICALS AND APPLICATION METHODS

1 Evaluation of chemicals for mosquito control	81
2 Evaluation of chemicals for tsetse fly control	81
3 Evaluation of chemicals for locust control	81

4 Bird control — avicides 82
  5 Miscellaneous chemical work 82
  6 Aerial spraying of cotton in Swaziland 82
  7 Spray trials on cotton in the Gambia 84
  8 Pesticide application to crops in Thailand 84
  9 Tests of ultra-low volume sprayers 85
 10 Evaluation of the Evers and Wall Mk II Exhaust Nozzle Sprayer 85
 11 Hovercraft 86
 12 Methods for the detection of acaricide resistance in ticks 86
 13 NRACC sub-committee on Pesticide Application Overseas 87

 I ENVIRONMENTAL STUDIES OF PESTICIDES 88

V INFORMATION

A SCIENTIFIC INFORMATION AND LIBRARY SERVICE   93

B PUBLICATIONS   101

C INDUSTRIAL LIAISON   102

D ECONOMIC SALTATORIA   104

E INFORMATION LECTURES   104

VI TRAINING

A COURSES AND LECTURES   105

B VISITING RESEARCH WORKERS   109

APPENDICES

 1 STAFF OF THE CENTRE FOR OVERSEAS PEST RESEARCH   111
 2 PUBLICATIONS OF THE CENTRE FOR OVERSEAS PEST RESEARCH   115
 3 OVERSEAS VISITS BY COPR STAFF   121
 4 CONFERENCES AND MEETINGS ATTENDED BY COPR STAFF   130
 5 COMMITTEES, COPR REPRESENTATION IN 1975   134
 6 COPR CONSULTANTS   136
 7 VISITORS TO COPR   137
 8 GLOSSARY   140

I GENERAL REVIEW

International concern over the world food situation, remarked on in last year's report, was reflected in a number of developments and initiatives relevant to the work of COPR. The World Food Council held its meeting in Rome in June and announced a number of 'priority areas' in which action was needed to safeguard or increase food production and one of these was plant protection. Various UN Agencies, including FAO, WHO and UNDP responded by organising international conferences on various aspects of this problem and initiated important projects related to pest control in agriculture and public health, in many of which the Centre is involved.

In Britain, the Ministry of Overseas Development, the Centre's parent body, following its earlier declaration of greater support for the poorest countries and within those countries to the rural sector, organised a Commonwealth Ministerial Conference on 'Food Production and Rural Development' in London during March. These developments gave added point to the revision of the programme resource allocation of the Centre begun at the end of 1974, and commented on in last year's Annual Report, in which more emphasis was to be given to overseas work with a direct bearing on food crops and increased attention paid to follow-up, training and demonstration projects.

However, major shifts of programme emphasis in 1975 were made difficult because of the combination of forward commitments of manpower and resources and by the continuing demand for COPR services which remained at the same high level as in 1974. The alternative solutions were an increase in manpower or a cut in some research programmes; hopes of the former were raised by the agreement of the ODM Manpower Committee to increase the COPR complement by six extra posts as from 1 April 1976, but in the event these posts were 'frozen' early in 1976 in the general review of civil service manpower. Accordingly the Board of Directors reviewed the whole position in 1975, reduced certain research programmes and re-allocated some posts between Divisions so as to be able to increase the percentage of resources available for overseas work.

The dialogue with ODM Natural Resources Research Department on extra home-based manpower for overseas work thereby became more important but as this coincided with a decision by ODM to review the whole field of manpower in the natural resources field, this process has taken much longer than expected and was not resolved during the period of this report.

The Centre continued its involvement with UN Agencies, particularly FAO and WHO, on the development of various international projects in the pest control field and developed further links with some of the international agricultural research institutes funded by the Consultative Group for International Agricultural Research. Details of these projects are given in Section IIB below, and Section IIA outlines the bilateral project position.

The relevance of the Centre's programme to the totality of the Natural Resources programme by ODM was ensured by the work of the ODM/COPR Management Committee and also by that of the Natural Resources Advisers Co-ordinating Committee (the NRACC). The Director continued as Chairman of the NRACC Sub-committee on Pesticides Overseas and this committee, representative of a wide range of government, state-supported and academic research institutes, is carrying out increasingly useful work.

STAFF AND ACCOMMODATION

An unusual number of staff changes took place during 1975 and the Centre was unfortunate in losing the services of Dr P. M. Symmons, Assistant Director and Head of the Division of Ecology, who resigned the service in November to take up appointment as the first Director of the newly established Plague Locust Commission in Australia. Dr Symmons, well known for his work on locust biogeography, originally joined the Anti-Locust Research Centre, one of the original constituent units of COPR, in 1965, from the International Red Locust Control Service in central Africa and was responsible for many years for the work of the Desert Locust Information Service; he played a leading role in building up what is now the Centre's Division of Ecology and was particularly interested in the development of combined 'research and extension' projects, which are now becoming the standard for overseas work.

A certain measure of poetic justice may be noted in relation to Dr Symmons' new appointment, because Sir Boris Uvarov, former Director of the Anti-Locust Research Centre, retained by the Australian Government as consultant to advise on the future of locust research and control in that country, proposed as a basic requirement the formation of a Federal Locust Commission!

Another long serving member of staff to leave was Mrs E. Blaxter who in her 20 years' service worked in the Library in ALRC and eventually became Public Relations Officer of the Centre; she resigned for health reasons and the good wishes of the staff go to her in her retirement.

New staff welcomed to the Centre during 1975 included, in the Administration group, Mr C. F. G. Foss as Administrative Secretary and Miss D. L. Moors, his Deputy, who replaced Mr C. T. R. Gordon and Mr F. Scrivener respectively. In the Division of Ecology, Dr J. Murlis joined the tropical meteorology group, Dr M. R. K. Lambert the armyworm ecology group and Miss E. M. Paton the locust ecology group. In September Mr J. Farrington arrived as the long-awaited economist for the Centre. In the Biology Division, Dr P. Matthiessen joined the molluscicide group to work on the toxicity and environmental hazard aspects of pesticides in aquatic media.

It proved impossible in 1975 to alleviate to any great extent the severe accommodation problems of the Centre. There is not only an overall shortage of space, but much of what is available is unsatisfactory as regards both laboratory and office accommodation. Much effort has gone into a search for more space nearby and in the more efficient use of that presently available, but unless some new accommodation can be found soon the pressure on the staff will become severe. Continued progress was made, in concert with our sister units TPI and LRD, and the architects group of PSA/DOE in the planning of the proposed new accommodation for the ODM Scientific Units at Bramley, near Basingstoke. The Feasibility Study was completed and is now being examined by ODM and the Units and it is hoped that a final decision on this important development will be taken in 1976. However that may be, no move can be made for at least five years and hence this proposed relocation does not contribute to a solution of the present problems in College House.

OUTSTANDING EVENTS

The year was notable for the large number of meetings, conferences and exhibitions organised by the Centre. From 3–5 February the Centre was host to an international workshop on the effect of pesticides in the environment, which brought together experts from the international agencies, including FAO, WHO, UNDP, UNESCO, and from six developing countries. The meeting discussed the research needs in this area and issued a report which made recommendations on the type of programmes and work needed in this important field. This report, later published in the new COPR 'Proceedings' series, has been much in demand since it constitutes one of the few published 'guidelines' to future action. This meeting was the first of what it is hoped will be a series of international workshops and seminars organised by COPR, with the backing and financial support of ODM, which will bring together experts from developed and developing countries to discuss aspects of pest control which are relevant to agricultural development overseas.

From 18—20 February the Centre held a series of 'Open Days' when its work was on display to the public. A press conference on the first day resulted in much interest by the national press and included a TV broadcast. The public interest shown was such that queues for admission stretched into the streets and there was difficulty in coping with the large numbers of visitors, a total of 560 attending for the three days. The exhibitions were of a high standard and were so constructed as to be usable for future occasions.

They were in fact used again to good effect on 8 March when the Centre was open to visits by delegates to the Commonwealth Ministerial Meeting on Rural Development, mentioned previously; a large number of delegates came including four Commonwealth Ministers of Agriculture and useful discussions on overseas collaboration and possible technical co-operation were held informally during the course of the day.

In June the Commonwealth Entomological Conference was held in London; the Director was a member of the UK delegation and also attended the CAB Review Conference held subsequently. Delegates from the Entomology Conference also visited the Centre for discussions with members of staff.

In early August the Centre was the venue for a UNDP/FAO meeting on plant protection training, which again emphasised the importance of this neglected aspect of pest control in relation to developing countries and their needs.

One of the most important events during the year was the premiere of the new film about the work of COPR, entitled 'Health and Harvest'. It had been decided in 1974 to make a film, suitable for showing both in the UK and overseas, about the work of the Centre in developing countries which would emphasise the practical application and importance of our research, information and training programmes. The film was made by Mr Stephen Boyland, a professional photographer, and Dr Graham Mitchell of the Division of Biology; they filmed research both in the UK and overseas and the film shows scientists from COPR and developing countries working together both in the laboratory and field.

A distinguished audience of Ambassadors and High Commissioners of countries with which COPR works and scientists and academics attended the first showing of the film at the Commonwealth Institute on 10 November, when it was introduced by Mr John Grant, the Parliamentary Under-Secretary of the Ministry of Overseas Development. The film will be made in several languages and is available for public showing at home and overseas and already to date some 20 requests for showings have been made.

The year of meetings came to a close on 22 December when the Centre gave a reception to mark the 21st birthday of its quarterly journal *Pest Articles and News Summaries* — or *PANS* as it is universally known. In recent years *PANS* has expanded both in content and readership and now reaches some 177 overseas territories; it is not a research journal but is concerned with practical problems of plant protection in developing countries. Its importance is shown by the fact that it came tenth in the world list of 'most cited agricultural journals' even in competition with commercial journals. Consideration is now being given to wider language coverage of *PANS* which is expected to increase its circulation even more.

As regards events during the year on the technical side, it was disappointing not to be able to accomplish more in training and demonstration; one of the main reasons for this was that the post of Training Officer, essential to the expanded programme was 'frozen' by ODM as mentioned above. Nevertheless some important work was carried out, particularly in the Sahel where a team of three COPR scientists working through the Office of Sahelian Relief Operations in FAO carried out a combined survey/training mission in several countries of the region in collaboration with the multi-country Organisation Commune de Lutte Antiacridienne et Lutte Antiaviaire (OCLALAV) (see Section IVE).

One of the major problems in agriculture in developing countries is rodent control, which is rapidly becoming the most serious pest problem in several areas. The large number of requests for technical co-operation in this field, which far exceeded the ODM response capacity, prompted the Centre to propose to the Ministry of Agriculture, Fisheries and Food a co-operative programme for overseas rodent research and control based on the expertise and resources of the Rodent Research and Control Division of the MAFF Pest Infestation Control Laboratory. These discussions, held on the ODM side under the aegis of the Natural Resources Advisers Co-ordinating Committee, are still continuing but useful progress has been made towards defining a basis of co-operation.

As indicated in last year's report, ODM had expressed readiness to participate in the new international campaigns against African animal trypanosomiasis and it was agreed that increased capacity for work in the control of the vector tsetse fly should be based on COPR. Unfortunately once again manpower difficulties have prevented recruitment of the necessary staff, which has slowed the build-up of extra capacity, but because of the importance of this work it is intended to utilise existing posts and recruitment will begin early in 1976. Meanwhile, Mr C. W. Lee, Scientific Secretary, in co-operation with Dr A. T. Jordan, Director of the ODM supported Langford Tsetse Laboratory at Bristol University is co-ordinating the Centre's tsetse fly work. The most important job in hand at present is a contract with FAO for production of a report on tsetse insecticides and application techniques with recommendations for future research, to be utilised in the FAO 5-year pre-project activity programme (see Section IIA).

COPR continues to co-operate closely with the WHO Onchocerciasis Control Project and had a team in the field operations area working on the effect of weather systems on the migration of the vector fly, *Simulium damnosum*.

The Centre is also involved in the WHO Tsetse and Trypanosomiasis Project and is discussing with the Government of Botswana, UNDP and FAO the possibility of mounting a project on the environmental effects of aerial spraying of endosulphan for tsetse fly control in the Okavango Delta.

Fig. 1
The Minister for Agriculture from India (left) discussing pest control problems with the Director, COPR, Dr P. T. Haskell during the 1975 Open Day Exhibition.

Fig. 2. Visitors from Swaziland taking an interest in ultra-low volume spraying during Open Day.

Collaboration between the Centre and some of the CGIAR international agricultural research institutes continues to flourish. As indicated in last year's report, IITA wished to expand and develop the present pesticides residue project (see Section IV I) into a project on pest control and the effect of pesticides in multiple cropping systems. Arrangements for this, which will involve the IITA taking the project, along with certain COPR staff, on its core budget, have nearly been completed and the new project should start in March 1976. Further collaboration with ICRISAT over stem borer research is being developed and discussions are taking place with IRRI on collaborative research on rice pests.

Finally, two important projects are under discussion for bi-lateral assistance. One of these is an integrated pest management project for cotton pests in Pakistan, which arose partly as a result of the FAO/UNDP Conference on Cotton Pest Control held in Karachi in October; it is intended that if this project goes forward it will be regarded as part of the FAO Global Programme of Integrated Pest Control.

The second scheme, now under discussion with the Environmental Protection Council of Ghana, is for a Biocide Testing Project; the objective of this is to provide facilities for the testing of acute and chronic effects of insecticides, molluscicides, herbicides, avicides and other agricultural chemicals in aquatic ecosystems. It is particularly relevant to environmental problems raised by the WHO Onchocerciasis Control Project in the Volta Lake and river system in Ghana and also on the use of molluscicides to control schistosomiasis in that area.

From this brief general review, which is expanded in the following sections of the report, it can again be seen that the Centre's staff and facilities were stretched to the limit during 1975; if the manpower situation does not improve in 1976 then it will be necessary to run down or stop certain UK based research and development projects in order to maintain work overseas at the present level.

HONOURS AND AWARDS

Dr Haskell was appointed a Companion of the Order of St Michael and St George in the New Years Honours List.

Dr Rainey was appointed a Fellow of the Royal Society in June 1975 for his work on insect migration, the citation being that he is 'distinguished for his studies of the relationships between weather patterns, migrations and breeding of the Desert Locust and the application of these studies to the control of this pest.'

II INTERNATIONAL CO-OPERATION

A UNITED NATIONS AGENCIES

UNITED NATIONS DEVELOPMENT PROGRAMME

A cooperative programme with the UNDP *Locusta* Project was carried out in Mali in October and November by the COPR Radar Team (Dr J. Riley, Dr D. Reynolds and Mr A. D. Smith) assisted by Dr J. Murliss and Mr M. Richie. On the Niger flood-plain they tested a new spinning-dipole, vertical-looking radar system and also studied the movement of acridids in the northern areas.

UNITED NATIONS ENVIRONMENTAL PROGRAMME

In January Dr W. A. Sands attended a meeting of UNEP representatives with ICIPE staff and Directors of Research (Termite Group) in Nairobi to discuss possible support for an extended programme of termite ecology. UNEP was interested in such a project and Dr Sands prepared an extensive planning paper setting out the research programmes, staff requirements, accommodation and equipment with 5-year budgets. A decision from UNEP is awaited.

FOOD AND AGRICULTURE ORGANIZATION
Plant Production and Protection Division
Desert Locust Control Committee

Mr Ashall and Mr Roffey attended the 19th Session of the Desert Locust Control Committee. The main subjects of direct concern to the Centre which were discussed were the progress of the UNDP Training Project on Crop Pest Control with special reference to Desert Locust Control, the Progress of the FAO/SIDA Project and the use of satellite application techniques to improve Desert Locust survey and control. COPR continued to maintain its record of Desert Locust incidence and to prepare six-monthly summaries of the Desert Locust situation for the FAO Locust newsletter.

As in past years COPR assisted the Committee in a number of activities.

Mr Roffey was invited by the FAO Locust Control Office to prepare a plan to improve Desert Locust survey and control by locating potential Desert Locust breeding sites using satellite observation techniques. Following visits to Rome for discussion with Dr J. A. Howard, Senior Officer of the FAO Remote Sensing Unit and to Washington D.C. and Houston for discussions with officials of the U.S. National Aeronautics and Space Administration and the National Oceanic and Atmospheric Administration a project plan was prepared which was discussed at a technical consultation convened in Rome and approved by the 19th Session of the DLCC.

The Centre continued to be involved with the FAO/UNDP Inter-regional Training Project and a mid-term review was held at the Centre in London in August.
Mr G. B. Popov assisted two local courses under the Project, both in French. He was co-ordinator/organiser for a course (on crop pest control with special reference to Desert Locust control and research) in Dakar in February and March and lectured to this and a similar course in Algiers in November.

Dr N. D. Jago spent three months in Ethiopia late in the year assisting Mr T. Crowe at the Institute of Agricultural Research, Holetta with the FAO programme in that country.

Entomology Working Group on Crop Losses

Mr P. T. Walker again worked as consultant and adviser to the crop loss group, both in preparation of supplements to the FAO Manual of Crop Loss Assessment and in planning workshops on pest and crop loss assessment in pest management.

A very successful workshop was held at the University of Bari, Italy, from 12 to 19 April. It was attended by scientists from 15 countries, and Mr Walker was flown from Kenya to lecture on pest assessment, methods of carrying out crop loss trials, and the relationship between pests and loss. A short period was spent in FAO in Rome afterwards with the crop loss group. There is now a wider interest in many countries in measuring the losses of food and cash crops to pests and diseases because of their importance in food supply and trade.

Animal Production and Health Division

Dr A. B. Hadaway and Mr E. G. Harris attended the FAO Expert Consultation on Research on Tick-borne Diseases and their Vectors in Rome on 6–8 May and took part in the First Session of the Joint FAO/Industry Task Force on Ticks and Tick-borne Disease Control that followed. One outcome was that COPR agreed to assist FAO in the development of a Test Method and Kit for the Global Acaricide Resistance Monitoring Programme. Work is proceeding at the Chemical Control Division on the comparison of different methods and on the effects of various modifications of techniques and the conditions of testing to be employed.

In October, COPR at the request of FAO undertook the preparation of a report on insecticides and application equipment for tsetse control. The report is to consist of a review of information on insecticide formulations and equipment in addition to the research requirements necessary to improve the technology of tsetse control in the future. Drafts are being prepared by Dr A. B. Hadaway and Mr F. Barlow on insecticides, Mr C. W. Lee and Mr J. Parker (WHO) on research and aerial application equipment and Dr W. R. Wooff (University of Salford) on ground application equipment. The document which is being prepared jointly by FAO, COPR and WHO will be submitted to the next meeting of the FAO/Industry Task Force on Tsetse Control in 1976.

Office of Sahelian Relief Operations

Mr G. B. Popov was a member of OSRO delegation at the CILSS/OCLALAV meeting held at Oagadougou to discuss the requirements of the national and regional plant protection services in the Sahelian countries for 1976/77 campaigns. These were formulated and submitted to OSRO for communication to potential donors.

Mr Popov was also in the field in the Sahel from mid-May to mid-December to serve as senior co-ordinator in the emergency campaign for the protection of subsistence crops against ravages of insect crop pests in the Sahelian countries. He was assisted by Dr M. R. K. Lambert and Miss P. McAleer; together they constituted the COPR input towards the Campaign. The team's activities covered many fields including (a) liaison with OSRO/FAO, the donors (e.g. USAID, CIDA, FAC), the regional pest control organisations (OCLALAV and OICMA) and the plant protection services of the Sahelian countries; (b) Training (and demonstration) of technical personnel of crop protection services and of farmers; (c) assistance with planning and execution of surveys and control operations; (d) special monitoring surveys/ studies to assess the crop pest situation and the level of damage; (e) studies on life cycles of some crop pests, and (f) reporting and forecasting.

WORLD HEALTH ORGANIZATION

Vector Biology and Control Division

Mr C. W. Lee has maintained close liaison with WHO through the Division of Vector Biology and Control particularly with the development of equipment to apply insecticides for the control of pests of public health importance. In February, Mr J. Parker (WHO Scientist/Engineer) visited the Centre and accompanied by Mr Lee attended meetings at OSMC with Dr Matthews (Officer-in-Charge) and spraying machinery manufacturers to discuss the development of a prototype for ULV application of residual insecticides for mosquito control in dwellings. As a result of these meetings, a prototype machine has been constructed and tested at OSMC and this will shortly be available for field tests by WHO (see Section III).

In February, Mr Lee visited the Director and staff of the Division of Vector Biology and Control to discuss the preparation of a report on aerial application equipment and to draw up a co-ordinated research programme aimed at improving the technology of tsetse control. These reports are being edited and will form part of the document to be published in 1976 by FAO.

The Division of Chemical Control, COPR, continued to serve as a WHO Collaborating Centre for the Evaluation and Testing of Insecticides. Candidate insecticides and formulations are assessed in laboratory tests for contact and residual toxicity to adult mosquitoes and tsetse flies.

WHO/IBRD Onchocerciasis Control Project

A contractual service agreement was signed in March between WHO and COPR, by the terms of which it was arranged for Dr J. I. Magor and Dr L. J. Rosenberg to spend 9 months and Mr D. E. Pedgley 2 months in the Programme Area studying the weather situations associated with changes in the distribution of *Simulium damnosum* to assess the probability of major redistributions arising from windborne movements. Short visits to the area were made also by Dr D. R. Reynolds and Mr M. J. Hodson of COPR to help decide what future entomological studies should be proposed. A report of the biogeographical studies and proposals for future work were submitted to WHO in November. Reinvasion in 1975 within the area being controlled emphasised the importance of understanding the movements of the flies.

Malaria and Communicable Diseases Division

A WHO research grant was received again this year to support the provision of a confidential bioassay service for molluscicides and for work on their mode of action. The grant also permitted further study of the sub-lethal effects of molluscicides and other pesticides on the tropical food fish *Sarotherodon mossambicus*. Particular attention was given to the uptake and excretion of the molluscicide Frescon by fish. Work was also begun on a study of the organo-phosphorus larvicide Abate, which is being used extensively now by WHO for *Simulium* control in the West African onchocerciasis scheme.

WORLD METEOROLOGICAL ORGANIZATION

As a Rapporteur to WHO, Mr D. E. Pedgley has started work on a draft report on the take off, movement, dispersion, deposition and survival of organisms in the air. The aim is to outline the nature and behaviour of atmospheric disturbances on various scales and to discuss how they influence the way organisms get into and out of the air and how they are moved around.

B INTERNATIONAL ORGANIZATIONS

ASSOCIATION D'ACRIDOLOGIE, PARIS

Miss J. M. Child visited Dr F. O. Albrecht in Paris in June to discuss the production of the Association's journal *Acrida*, the English language papers of which she edits. Mrs J. S. Ridout of SILS continued to prepare Acridological Abstracts for publication in *Acrida*.

ASSOCIATION FOR THE ADVANCEMENT OF AGRICULTURAL SCIENCES IN AFRICA

The Association continued to make progress, as noted in last year's report, under its new Secretary-General, Dr L. K. Opeke. A highly successful Second General Conference of the Association was held in Dakar in March. Dr B. Steele, whose secondment to the Association as Publications Adviser continued until May, also assisted in the organisation of this Conference.

COMITE EUROPEEN DE NORMALISATION

Mr R. M. C. Williams of the Termite Unit is cooperating with this organization through the British Standards Institution in discussing details of the proposed European wood preservative termite testing standards for its Working Group 38 (Wood preservative tests).

DESERT LOCUST CONTROL ORGANIZATION FOR EASTERN AFRICA

COPR continued its close cooperation with the Organization. Much attention was given to future plans to extend DLCOEA's interests to Armyworm forecasting. In this connection, Mr C. Ashall held discussions with the Director in January and attended the Technical Committee as a consultant in October. ODM has also renewed the secondment of Mr P. J. Kercher as Technical Adviser to the Director.

EAST AFRICAN AGRICULTURE AND FORESTRY RESEARCH ORGANIZATION

EAAFRO is much involved, together with DLCOEA in setting up the Eastern African Armyworm forecasting network and this was the main area of cooperation with the Organization in 1976. Mr C. Ashall visited the Director for discussions and Miss E. Betts maintained her technical liaison with EAAFRO officers.

INTERNATIONAL CENTRE FOR INSECT PHYSIOLOGY AND ECOLOGY

Dr P. T. Haskell is a member of the Board of Governors of ICIPE and chaired the meeting of the Executive Committee of the Governing Board in September. He and Dr W. A. Sands also attended the Research Conference in Nairobi in June where also there was a discussion about the setting up of the EAAFRO/DLCOEA Armyworm project.

Dr W. A. Sands is Director of Research for ICIPE in charge of termite ecology. He supervises the work of Dr J. P. E. C. Darlington who arrived in Kenya in January after preliminary training in termite taxonomy with the COPR Termite Unit in London, followed by a visit to the Centre's Mokwa Termite Ecology project in Nigeria for field experience.

Dr Sands' first visit to ICIPE in January saw the establishment of the ecological programme to study the role of *Macrotermes subhyalinus* as a grass consumer in a semi-arid ecosystem. Dr Darlington is studying the internal economy of the colony which necessitates careful mound sampling.

During the January visit, Dr Sands also cooperated with Dr Darlington and Dr D. Pomeroy of Kenyatta College in a short ecological project on the settlement of reproductive pairs of *Hodotermes mossambicus*, another grass-feeding termite. A short paper has been prepared under joint authorship.

On his second visit to ICIPE in June, Dr Sands accompanied Dr Haskell to the Annual Research Conference and Policy Advisory Committee meetings. This was the year for the full review of the termite programme with individual presentations by Directors of Research and Research Scientists.

This visit to Kenya coincided with the arrival of the second termite ecologist of the group, Dr M. Lepage, who is working under Dr Sands on the external economy of *M. subhyalinus*. While in Nairobi on both visits, Dr Sands held detailed discussions of work in progress with all ICIPE termite group scientists.

There is also close cooperation with ICIPE by the staff of the Centre concerned with armyworm problems and in January a joint Armyworm Workshop was held in Nairobi.

INTERNATIONAL CROPS RESEARCH INSTITUTE FOR THE SEMI-ARID TROPICS

Dr J. C. Davies continued his five-year secondment to ICRISAT as Chief Entomologist during the year. He is currently in charge of the entomological work on the five crops of direct concern to the Institute — sorghum, pearl millet, chickpea, pigeon pea and groundnuts. An extensive programme of sampling, damage assessment and pest biology studies is now underway. This secondment also provides a channel for cooperation with COPR staff in Britain. Dr R. E. Roome has, for example assisted with Dr B. Nesbitt of the Tropical Products Institute in work on the pheromone of the sorghum stem borer (*Chilo partellus*). This pheromone is now being tested in the field by Dr Davies.

INTERNATIONAL RED LOCUST CONTROL ORGANIZATION FOR CENTRAL AND SOUTHERN AFRICA

Mr C. Ashall continues to maintain the close links between the Centre and IRLCOCSA in his capacity as technical consultant to the Council. In the field Mr J. Tunstall has continued his twice yearly Red Locust surveys by land and air in Zambia and Tanzania.

INTERNATIONAL RICE RESEARCH INSTITUTE

The Institute has a project at its Bangladesh outstation under Dr D. Catling, an ODM officer. Dr R. E. Roome has cooperated with Dr Catling and with Dr B. Nesbitt of TPI on work on the Rice Stem Borer (*Chilo suppressalis*).

ORGANISATION COMMUNE DE LUTTE ANTIACRIDIENNE ET DE LUTTE ANTIAVIAIRE

The main area of cooperation with OCLALAV during 1975 was in the organisation of Mr G. B. Popov and other COPR staffs' activities in the crop protection operations in the Sahel (see Section IVE(1)). Since the main pests involved were grasshoppers it was appropriate that OCLALAV should be the main local cooperating organization in the field.

ORGANIZATION FOR ECONOMIC COOPERATION AND DEVELOPMENT

Dr P. Rosen took up his assignment as Scientific Secretary/Coordinator of the OECD Steering Group on Pest Control for Small Farmer Food Crops, on 1 June in office space provided by the Centre. The Chairman of this international Group is Dr P. T. Haskell, Director, COPR and Dr Rosen's assignment is supported by ODM research funds.

The Group is charged with the preparation of a report the main aims of which are:

1. To identify the factors responsible for low food crop production by small farmers in developing countries, in particular the losses caused by pests.

2. To review the activities of UN agencies, national and international institutes and others, in the general area of crop protection in developing countries and particularly at the small farmer level.

3. To identify the gaps in knowledge and to define needs for additional research necessary for improving crop protection.

4. To propose specific projects for research/demonstration/training that will provide guidelines for crop protection in developing countries.

Dr Rosen has already collected much material for the report. Apart from regular Group meetings, Dr Rosen has held discussions with individual Group members and he has visited FAO in Rome, University of Wageningen, Holland, OECD in Paris, WHO in Geneva, UNDP and the World Bank in the USA and the International Research Centre, Canada, in addition to scientific institutes in the U.K.

COPR scientific staff have given considerable help in the compilation of the report and Dr Rosen receives assistance from the staff of the library and information services.

ORGANISATION INTERNATIONALE CONTRE LE CRIQUET MIGRATEUR AFRICAIN

In July, Dr N. D. Jago attended the meeting of the Council of OICMA in Cameroon as representative of COPR. Dr Jago gave advice on the recording of locust populations by the pictorial–graphical technique and this was adopted by the Council. Discussions were also held with the Director-General of OICMA on improving the assistance given by COPR in identifying and curating insect collections, reorganising records and training staff in grasshopper recognition.

In October and November the COPR radar team (Dr J. Riley, Dr D. Reynolds and Mr A. D. Smith), together with Dr J. Murlis and Mr M. Richie worked in the Niger flood-plain in Mali studying movement of Acridids in the northern areas and carrying out trials of a new spinning-dipole, vertical-looking radar. OICMA acted as host organization for the visit.

A long paper by Dr R. A. Farrow on the ecology of the African Migratory Locust, resulting from his studies with OICMA from 1963 to 1970, was published as No. 11 of OICMA's non-periodic journal *Locusta*. COPR gave considerable assistance with this undertaking, not only arranging the shortening and editing of the paper but also supervising the printing in England and checking of proofs. Arrangements are in hand for the publication of a French version.

C INDIVIDUAL COUNTRIES

AUSTRALIA

Dr R. E. Roome, in cooperation with the Unit of Invertebrate Virology at Oxford, has assisted Dr K. Wardhaugh of CSIRO on *Heliothis* problems. Advice has been given on rearing the insects and samples of diseased *Heliothis* have been submitted to UIV Oxford for identification of the virus infection.

BANGLADESH

Following a request from the Government of Bangladesh, Mr C. Ashall and Mr P. T. Walker visited Bangladesh in November and December for discussions on various pest control problems, working in close collaboration with a USAID team engaged in a more general survey. Several important side issues, such as storage and pesticide disposal were examined and passed to the relevant experts, but the main area of cooperation proposed is in crop pest and loss assessment, both to improve the information on which pest control decisions are based and to provide data for forecasting and investigating pest outbreaks. A report has been written and the matter is under discussion.

BOTSWANA

In July, Mr C. W. Lee and Mr G. G. Pope visited Botswana to monitor fixed-wing airspraying operations being performed by the Department of Veterinary Services and Tsetse Control against tsetse in the Okavango Swamps. The opportunity was taken to assess two new insecticide formulations, a synergised pyrethrin and a ULV formulation of endosulphan. Only the latter gave promising reductions in tsetse population.

During the same visit, Mr Pope carried out successful trials of the use of hovercraft in swampy conditions both for communication and spray application.

Liaison between the Division of Agricultural Research and the Unit of Invertebrate Virology at Oxford has been established through Dr R. E. Roome. Studies of *Heliothis* are continuing in Botswana and samples for virus assessment have been sent to Oxford.

The report of the study on damage caused by doves to sorghum by Mr N. S. Irving and Mr J. S. S. Beesley made in 1974 was completed in 1975. Its recommendations were considered and it was agreed that ODM should continue to support the work on birds in Botswana by seconding Mr Beesley to the ODM Dryland Farming Research Project already in operation.

UNITED REPUBLIC OF CAMEROUN

During a visit to West Africa Dr N. Jago established liaison with the plant protection services of Cameroun and visited several research institutes in the country. A number of professional contacts were made with the staff of the Ecole Nationale Superieure Agronomique at Yaoundé.

Fig. 3 Dr. G. A. Mitchell inspecting a banana weevil trap beneath the cut end of a banana 'stem'. St. Lucia.

Fig. 4

Spraying a mixed crop of cocoa and bananas in Grenada.

CARIBBEAN ISLANDS

Antigua

Mr N. S. Irving, a COPR entomologist, is stationed in Antigua as one of the team of three entomologists for the Caribbean area. He has special responsibility for advice on the pests of Sea Island Cotton and their control.

A survey of the timing and extent of damage by the Sweet Potato Weevil was carried out in January by Mr P. T. Walker. Recommendations were made for the replacement of the present system of control by organochlorine insecticides by the use of cultural methods and resistant varieties of potato.

Barbados

Mr W. R. Ingram is the COPR entomologist stationed in Barbados and is Regional Adviser to the British Development Division in the Caribbean on crop protection matters. He is working on the entomology of the Sea Island Cotton crop in the Barbados, evolving scouting techniques, testing insecticides and sprayers and training local staff.

French Antilles: Martinique and Guadeloupe

The Plant Protection Organisation in Martinique was visited by Mr P. T. Walker, on his tour of the West Indies, to exchange information and make contacts on crop pest control with French counterparts. The agricultural research centre in Guadeloupe was also visited and information on a wide range of subjects, including biological control, cotton, fisheries, storage and plant diseases, was exchanged and subsequently passed to relevant British contacts.

Grenada

Mr Ingram has paid two visits to Grenada during the year, the first to assess pest and disease problems particularly in coconuts and cocoa and the second to observe cocoa spraying.

Puerto Rico

A short visit was paid by Mr Walker to the University Faculty of Agriculture for discussions on Sweet Potato Weevil attack and control. Trials on crop resistance and insecticides were inspected.

Windward Islands

The third member of the COPR Caribbean entomology team is Dr G. Mitchell who is stationed in the Windward Islands. He is seconded to the WINBAN Research Station at Roseau and is concerned mainly with investigation of the banana pest problems of the Windward Islands.

Whilst in the West Indies, Mr P. T. Walker paid a visit to WINBAN to discuss measurement of crop losses in bananas and the application of insecticide granules.

CENTRAL AFRICAN REPUBLIC

A large collection of termites (155 vials containing 56 species) made by Prof Dr Gunther Becker of the Bundesanstalt für Materialprüfung in Berlin from the Central African Republic has been identified by Mr S. Bacchus.

COSTA RICA

Dr A. B. S. King has been seconded through ODM to the Government of Costa Rica since June 1975. He is conducting research at the Centro Agronomico Tropical de Investigacion y Enseñanza (CATIE) at Turrialba on the major pests affecting the production of beans, maize, rice, cassava and sweet potato.

Costa Rica was also one of the many countries which assisted Miss S. C. A. Cook when she undertook a tour to collect information for the revision of *PANS* Manual No. 1, *'Pest Control in Bananas'*.

EGYPT

Miss S. Green has given advice to colleagues at the Plant Protection Research Institute at Dokki on the analysis of locust morphometric data.

The Egyptian Ministry of Health organised a valuable International Conference on Schistosomiasis in October. At this Conference Dr J. Duncan of COPR represented ODM.

ETHIOPIA

Dr N. D. Jago visited Ethiopia for ten weeks from mid-September at the invitation of the FAO and the Ministry of Agriculture, Addis Ababa. The aim was to organise a reference collection of Ethiopian grasshoppers and to make a survey of different areas to assess grasshopper damage. Over 4000 specimens were collected from some 90 localities. Results of this study are being completed, but many groups are in need of revision before correct determination can be made and over 30 species were added to the Ethiopian fauna.

Important recommendations were made on future research and control of 'degeza', a new bush-cricket (sub. F. Dectianae), which infests teff in the Blue Nile Gorge region.

Liaison with the Institute of Agricultural Research at Holetta concerning the African armyworm was continued by Miss E. Betts. During the year two warnings of possible moth invasion and subsequent larval infestation were given.

THE GAMBIA

Mr P. H. Goll is seconded to the Medical Research Council's outstation at Fajara, Banjul. He has started a four and a half year assignment to study the possible control of *Schistosoma haematobium* in seasonal pools, the longevity of the female bilharzia parasite in humans and the possible spread of transmission of bilharzia with increased swamp-rice cultivation.

WEST GERMANY

The termite group of the Biology Division donated a colony of the dry-wood building-pest termite *Cryptotermes havilandi* to Prof Dr G. Becker of the Bundesanstalt für Materialprüfung, Berlin, for evaluation of its potential for materials tests.

Fig. 5 Kenya, May 1975. The COPR team, Mr. P. T. Walker, Mr. M. J. Hodson and Mr. B. W. Bettany, loading up after giving a lecture and film on maize stem borer at the District Office, Vihiga.

Fig. 6 Kenya, April 1975. Mr. P. T. Walker lecturing and giving a field demonstration on maize stem borer.

GHANA

A two-week visit was made by Mr R. M. C. Williams of the Biology Division and Dr M. B. Usher of the University of York to the Building & Road Research Institute, Kumasi, to advise on current and projected termite materials-testing work. Proposals for 7 testing projects were accepted by the Director, BRRI, together with a suggestion that Mr J. K. Ocloo, in charge of the testing unit at BRRI, should visit COPR and other institutes in the UK for further training during 1976. This visit to Ghana was requested as further aid following the termination of COPR's recent technical aid scheme there, which set up a unit for testing materials against termite attack. Several UK industrial firms have been advised of the testing facilities at BRRI.

Mr Williams and Dr W. A. Sands have also continued contact in an advisory capacity with the Forest Products Research Institute, Kumasi.

Contacts were continued with the Ghanaian Environmental Protection Council in connection with the forthcoming Ghana Biocide Project in which the ecological side effects of tropical pesticides will be monitored in artificial fish ponds to be built near Accra.

GREECE

At the invitation of the Greek Ministry of Agriculture, studies on the use of pheromones for the control of *Spodoptera littoralis* were initiated in Crete by Dr D. G. Campion. The results of these preliminary trials were very promising and plans were made for more extensive work in 1976.

Mr D. J. McKinley also spent five weeks in Crete carrying out field work on the field status of the nuclear polyhedrosis virus of *S. littoralis* as a preparation for limited field trials next year. Future plans for this work have been discussed with Dr Mourikis and Dr Yamvrios at the Benaki Phytopathological Institute of Athens. Both programmes are carried out with close cooperation from the Agricultural authorities in Crete.

KENYA

In co-operation with the Kenya Ministry of Agriculture and the ODM Maize Agronomy Research Project in Kitale, Mr P. T. Walker, together with Mr M. J. Hodson and Mr B. W. Bettany, made an extension tour of the main maize growing areas of Kenya in April and May. A film on Maize Stem Borer made by Mr Bettany and Mr Hodson was shown, and lectures were given to technical staff and farmers on the pest and its control. The tour was very well received. Details are given in Section VI B. Some assistance has also been given with pests of oil-seed crops.

A joint COPR/ICIPE workshop on armyworm was held in Nairobi in January. There has been considerable cooperation on armyworm problems with the East African workers throughout the period, particularly with Mr P. Odiyo of the Armyworm Forecasting Service, who has also had discussions with Mr D. J. McKinley on the progress of current work on armyworm virus.

Dr W. A. Sands has made liaison visits to a number of Nairobi-based organizations when visiting Kenya for the ICIPE Directors of Research meeting. Cooperative contacts were established between the UNDP/FAO Game and Range Management Project staff and the ICIPE termite team. An arrangement for similar cooperation was made with the Environmental Monitoring Programme, a Canadian-funded project studying climate, soils and vegetation over a wide area of East Africa.

Dr J. Sale is coordinator of research at the Tsavo Research Centre. Dr Sands advised him and a research student, Mr R. Buxton, on the development of the programme of study on the decomposition of plant residues by termites. The student was also given a period of training in termite taxonomy at the COPR Termite Group in London before starting work in Kenya. Dr D. Pomeroy of Kenyatta College, Nairobi University, was advised on research on the development and erosion of *Macrotermes* mounds and a seminar was given to staff and students at the College on the distribution of East African termites.

MALAWI

Mr P. T. Walker has advised research workers in Malawi about insecticide use in maize for the control of stem borers and the prevention of residues in other crops following the treatment of maize.

The programme of investigations of the use of pheromones to control the Red Bollworm of cotton (*Diparopsis castanea*) undertaken by Mr R. J. Marks for ODM continued throughout the year. COPR is responsible for the technical direction of the project.

MALAYSIA

Mr R. A. Steedman visited Malaysia during September to study the habitats of *Valanga nigricomis*. In conjunction with staff from FAO and the Federal Land Development Authority, several infestations were visited, mainly in cocoa and oil palm plantations.

MALTA

The three year ODM research scheme on the control of farm flies, in which COPR collaborated with the Department of Agriculture and the Royal University of Malta, was concluded in September 1974. Papers on various aspects of the work have been prepared and will be published during 1976.

The results of the Gozo field trial indicated that, although high fly mortality was observed and insecticidal deposits on walls remained toxic throughout the season, effective control of horseflies and stableflies was not achieved. It is believed that a residual insecticide spraying campaign could only be successful if extended to include the majority of animal units and supplemented by a comprehensive programme of farm hygiene improvement.

MAURITIUS

The blood sucking fly *Stomoxys nigra* is a serious pest of cattle necessitating the housing of animals throughout the year and imposing a restraint on the development of the livestock industry. Mr J. E. H. Grose undertook a four week consultancy in November/December to assess current research programmes and to advise on future strategies for controlling or eradicating the fly.

A request was received from the Mauritius Government also for advice on the control of the Natal fruit fly. Miss E. J. Luard paid a two weeks' visit to assess the economic importance of this pest, but no technical cooperation project has yet been agreed as a result of her report.

MOROCCO

At the request of the Moroccan Ministry of Agriculture several pheromone traps and a supply of the pheromone of the Egyptian Cotton Leafworm were sent for local trials. If the initial experiment is successful it is proposed to set up a network of 50 such traps with the help and cooperation of COPR and to maintain these traps for a year in 1976.

NIGERIA

Mrs K. A. and Mr D. R. Johnstone have collaborated with Mr J. G. Quinn, Horticulturalist with the Institute for Agricultural Research at Samaru in the production of a report on fungicide spray trials on tomatoes carried out there in October 1974 and in planning future developments of such applications.

Dr N. D. Jago has supplied a reference collection of Nigerian Acridids to the University of Ibadan.

The project for investigating *Zonocerus variegatus* continued and is reported in Section IV A of this report. Dr E. A. Bernays has cooperated with Dr W. Hodder of the Department of Zoology at the University of Ibadan who has been working on *Zonocerus* feeding. The change in preference for cassava has been elucidated.

Miss S. Green has assisted Dr A. Yondeowi with the analysis of his data measuring the aggregation of the cocoa capsid.

PAKISTAN

Following discussions between the Director, Dr P. T. Haskell and representatives of the Pakistan Central Government on the possibility of UK technical co-operation with Pakistan on cotton pest control, Mr J. P. Tunstall paid a consultancy visit in December to identify and formulate such a project on the integrated control of cotton pests. Whilst emphasis is rightly placed by the Pakistan authorities on the development of an integrated control programme, the application of insecticides is at present providing the only available method of control. It is suspected, however, that the efficiency of insecticide use could be improved and that farmers are not at present obtaining the benefit they should expect from their control programme.

The UK technical cooperation project proposed is to examine in detail the insecticide requirement of an integrated control programme by means of a team of four scientists in the field in 1977. There are to be three main objectives: the development of an insecticidal spray programme for the small farmer, with special emphasis on the type of insecticide and timing and method of application by the ULV method with hand appliances; to assist the further development of aerial application of ULV sprays and to conduct in-service training and arrange overseas training for staff relevant to an integrated pest control programme.

SRI LANKA

Dr W. A. Sands was invited by the Tea Research Institute at Talawakele to visit Sri Lanka from 12 September to 6 October to study the problem of damage to tea by Kalotermitidae. In the low altitude estates, *Glyptotermes dilatatus* and *Neotermes greeni* do serious damage while at higher altitudes, *Postelectrotermes militaris* is the main pest. Whilst in Sri Lanka, Dr Sands visited many tea estates to assess the problem in the field. Recommendations were made for research on the biology and taxonomy of termites as well as proposals for research on the physiology of tea.

Seminars on termites as pests of crops and trees were given to staff at the TRI, Talawakele and to staff and students at Colombo University.

SUDAN

Dr J. Duncan visited the Gezira irrigated region to see recent developments in the control of bilharzia, particularly aerial application of molluscicides. Discussions were held on possible COPR inputs to this scheme but there has been no immediate result from these talks.

A warning of possible moth invasion and subsequent infestations of African armyworms in Southern Sudan was issued in early September.

SWAZILAND

A preliminary visit was made to the Commonwealth Development Corporation's Irrigation Scheme in Swaziland by Dr J. Duncan and Dr A. Fenwick of the London School of Hygiene and Tropical Medicine. Guidance was required on assessing the importance of the bilharzia problem and methods of dealing with it if necessary. It was recommended that one of COPR's biologists should be seconded to Swaziland for three years to work on bilharzia control.

SWITZERLAND

The termite group of the Biology Division donated a colony of the North Caribbean dry-wood termite *Neotermes jouteli* to Prof. M. Lüscher, University of Bern, for studies of the physiology of caste determination.

Fig. 7
Colony founding pair of dealate reproductives of *Glyptotermes dilatatus* in small chamber they have excavated in a dead pruning snag on a tea bush.

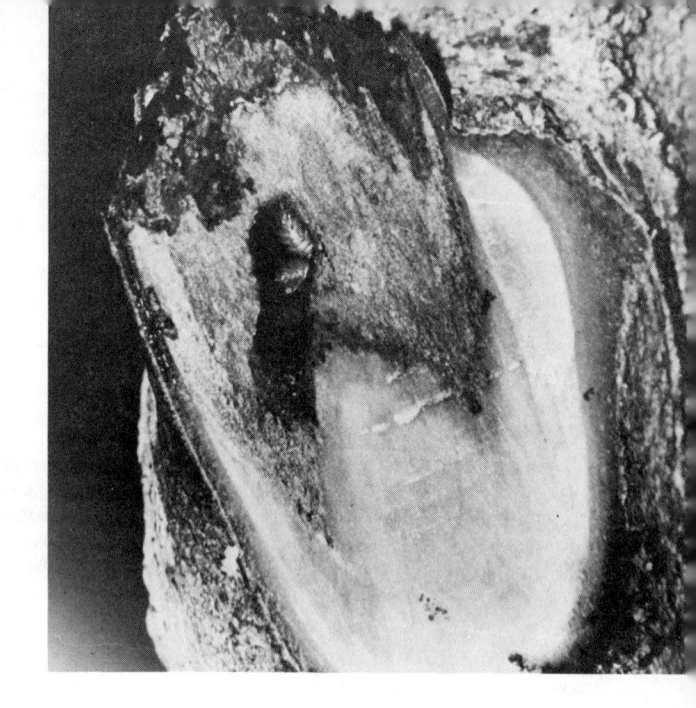

Fig. 8. *Postelectrotermes militaris* in galleries in up-country tea in Sri Lanka.

TANZANIA

An outbreak of *Spodoptera exempta* at Kilosa was investigated by Mr D. McKinley at the invitation of the Entomologist, Ilonga Research Station, Mr A. Mushi. Insect material was collected and subsequently found to contain virus.

An earlier survey of pests and losses in cereals in Tanzania by Mr P. T. Walker was followed up with a discussion on possible cooperation with the Faculty of Agriculture, Dar-es-Salaam in a pest and crop loss survey. Initial plans for this co-operation were developed.

Mr L. S. Flower carried out periodic monitoring of bran and hay samples for the presence of pp'DDT and γ BHC, fenitrothion and malathion from the Tsetse Research Project, Tanga, Tanzania.

THAILAND

Mr R. A. Steedman paid an 8-month visit to Thailand from January—September to assist in the establishment of a newly-formed Locust Control Research Centre. He was accompanied for short periods by Mrs A. Steedman and Mr R. J. Douthwaite and Mr J. Tunstall. The team considered and made recommendations on formal and in-service training as well as biogeographical and behavioural research, the results of which could be used to formulate a control strategy.

USA

The termite group of the Biology Division donated a colony of the introduced European subterranean termite *Reticulitermes flavipes santonensis* to Miss F. L. Carter of the USDA Southern Forest Experiment Station, Gulfport, Mississippi, for comparative studies with American species of that genus.

Mr Roffey visited Washington DC and Houston to discuss the application of NASA satellite imagery to the location of potential desert locust breeding sites. A programme plan incorporating use of LANDSAT imagery was prepared on behalf of FAO for the Desert Locust Control Committee.

YEMEN ARAB REPUBLIC

Assessments of the current African armyworm situation in eastern Africa were sent to Yemen Arab Republic on four occasions during the summer season, for the information of pest control organisations there following the very heavy and unexpected infestations of this pest the previous year.

III CO-OPERATION WITH UK BASED ORGANIZATIONS

ANIMAL VIRUS RESEARCH INSTITUTE

Mr D. E. Pedgley and Mr M. R. Tucker have assisted Dr R. F. Sellars in examining the possibility of windborne migration of virus-carrying midges in the spread of African Horse Sickness on several occasions in the past 50 years. There is circumstantial evidence that midges are carried downwind.

BRITISH MUSEUM (NATURAL HISTORY)

The Entomology Department of the British Museum (NH) has continued to provide accomodation for the COPR Termite Group, which takes complete responsibility for curating the national termite collection. Fewer specimens were added to the collections than last year. Mr R. T. Thompson, a member of the BM (NH) Coleoptera Section, was allowed to use the COPR 'Censor' Automatic Measuring equipment for a taxonomic project with an estimated saving of nine man-weeks work as compared with conventional methods.

The Acridid Taxonomy Group also works at the Museum under Dr N. D. Jago and makes its contribution to the curation of the collections. Considerable reorganisation work was undertaken by Mrs S. E. J. Storer. Following the publication of a review of the sub. f. Oxyinae by Mr D. Hollis this group of genera was completely rearranged and rehoused. The Euryphyminae, Calliptaminae and Pyrgomorphidae (totalling 167 drawers) were curated and re-labelled. Cabinet revision of the genera *Cataloipus* and *Heteracris* has been completed using mainly Museum material.

Mrs A. Steedman in cooperation with Dr Ragge has used the Museum's collection of *Ruspolia differens* to prepare a distribution map of this insect.

BRITISH STANDARDS INSTITUTION

COPR is represented on the Technical Committee PCC/1 on Common Names for Pesticides by F. Barlow of the Division of Chemical Control. Seven meetings were held during the year. The Third Supplement to British Standard 1831 has been published and addenda to ISO R1750 are being prepared.

Mr R. M. C. Williams of the Division of Biology termite group is a member of two BSI Technical Committees, WPC/2 on Classification of Wood Preservatives and WPC/10 on Testing of Wood Preservatives. WPC/10 met once during the year. Technical comments were made on a draft British and European Standard for WPC/10.

BUILDING RESEARCH ESTABLISHMENT, PRINCES RISBOROUGH LABORATORY

The agreement has continued between the termite group, Biology Division, and Mr J. M. Baker, head of the Biodeterioration Section of BREPRL, to partition materials tests. All tests of a standard kind are carried out at Princes Risborough, while those requiring special methods are done by COPR. During the year several industrial firms requesting tests were directed to apply to BREPRL.

Mr Baker is also one of the independent assessors of the work of the termite group, and reported on its scientific and technical content, working facilities etc in September 1975.

COMMONWEALTH INSTITUTE OF ENTOMOLOGY

The COPR Termite Group has for 25 years dealt with identifications and queries on termites received by CIE. This cooperation continued in 1975.

The Institute, and particularly its library, have been of considerable assistance in naming insects and tracing references for the staff of *PANS*.

COMMONWEALTH INSTITUTE OF HELMINTHOLOGY

The CIH continued to give advice to the staff of *PANS* on nematodes and their taxonomy during the year under review.

COMMONWEALTH MYCOLOGICAL INSTITUTE

Liaison has been maintained by Mr P. T. Walker with the ODM Plant Pathology and Forest Pathology Liaison Officers at CMI. He discussed various entomological problems with the former, Mr J. M. Waller and forest pest problems, particularly in Bangladesh with Dr I. A. M. Gibson.

The liaison officers also continued to give valuable support to *PANS*. CMI staff provided specialist advice on particular pathogens and their taxonomy. Miss S. C. A. Cook visited CMI several times to discuss papers, use the library and to clarify taxonomic points. She also attended the official opening of the new Culture Collection Laboratory in July by Lord Zuckerman.

GLASSHOUSE CROPS RESEARCH INSTITUTE

Mr P. T. Walker has given advice to the Institute on overseas and FAO contacts on new methods in hydroponics and also on possibilities in pest control with the thin film technique.

METEOROLOGICAL OFFICE

The Bracknell Meterological Office has, as in the past, continued to supply information on various projects. Mr J. B. McGinnigle provided meteorological data relative to pheromone trap catches in Cyprus which assisted in the preparation of a paper on this work. Mrs A. Steedman has used the Library and facilities of the Marine Division to collect extra data for her research project.

MINISTRY OF AGRICULTURE, FISHERIES AND FOOD

Mr F. Barlow has continued as a member of the Formulation Panel and Petroleum Oils Subcommittee of the Pesticides Analysis Advisory Committee of this organization. They met three times during the year. COPR has been taking part in collaborative tests on emulsion stability measurements and on the determination of oils and surfactants in miscible oils.

Two meetings of the Pesticides Analysis Advisory Committee, Granules Group of the Formulations Panel, were attended by Mr P. T. Walker. Specifications for granular pesticides were discussed, in particular particle size and number, rate of break-up, and abrasion.

OVERSEAS SPRAYING MACHINERY CENTRE

COPR maintains a close link with the OSMC particularly on research and development of spray application equipment for overseas situations. Recent collaboration between OSMC and the Centre has been concerned with a project to study ULV equipment and techniques for mosquito control. Work on the project has been mostly done at OSMC by a research student from grants provided by WHO and COPR. Several different types of equipment were assessed but the most promising device was a prototype ULV rotary atomiser fitted with a small fan manufactured by Micron Sprayers Ltd. Laboratory tests will soon be completed and these will be followed up by field tests in West Africa and in Thailand at the ODM Pesticide Application Research Unit, Bangkok.

Information sheets on pesticide application equipment designed to assist workers involved in crop protection and public health pest control overseas have recently been compiled by OSMC and COPR. These sheets contain an independent study of physical and mechanical tests performed on various types of apparatus which helps to determine the reliability and general performance of spraying equipment. The first series of information sheets have been published (*PANS* Vol 21.(4) Dec. 1975) and as new information becomes available, supplements will appear from time to time in *PANS*.

ROTHAMSTED EXPERIMENTAL STATION

Collaborative work has been carried out by Dr J. Murlis with Dr T. Lewis of Rothamsted on the orientation of the pear moth (*Cyclia nigricana*) within pheromone plumes and on the structure of the plumes. Mr B. W. Bettany made film of moths flying in plume. This work provided valuable design data for experiments on *Spodoptera littoralis* in Crete.

Dr L. R. Taylor has provided mapping facilities relative to pheromone trap catches in Cyprus and Miss S. M. Green has received assistance from the ODM statisticians at Rothamsted.

ROYAL BOTANIC GARDEN

Cooperation from two phytochemists at the Royal Botanic Garden, Dr T. Swain and Dr G. Cooper-Driver, has been very useful to Dr E. Bernays, particularly their advice on feeding inhibitors in locusts. A joint project on changes in the palatability of *Pteridium aquilinum* through the season was undertaken. Bioassay at COPR together with chemical assay at the Botanic Gardens has indicated the role of specific feeding inhibitors.

TROPICAL PRODUCTS INSTITUTE

The Molluscicide Unit of COPR has continued to be housed at the Tropical Products Institute. The Institute has acquired additional accomodation in the adjoining Thresh House and the Molluscicide Unit is now located there. Cooperation with the TPI Pesticides Section was established on a new project which has started to investigate the effects of the OP larvicide Abate on fish.

The Institute's Tropical Stored Products Research Centre has given advice to the Termite Unit on the safe use of methyl bromide for the fumigation of *Macrotermes* mounds and a member of their staff gave field guidance in Kenya.

There is active cooperation between the Centre's workers on pheromones and Dr Brenda Nesbitt of the Institute. She has recently extracted and identified the pheromones of *Chilo partellus* supplied from culture by Dr R. E. Roome. As a result the pheromones are now being synthesised and supplied to Dr J. C. Davies for testing during his secondment to ICRISAT.

UNIT OF INVERTEBRATE VIROLOGY

COPR has a collaborative project with the Natural Environment Research Council's Unit of Invertebrate Virology financed by ODM. In this project Mr D. J. McKinley has worked in close collaboration with the staff of this Unit on *Spodoptera* nuclear polyhedrosis.

Dr R. E. Roome has also undertaken informal liaison between the Unit and scientists working on *Heliothis* in Australia and Botswana to further the mutual interest in *Heliothis* virus infections.

UNIVERSITIES

COPR works closely with most universities, both in the field of training and cooperative research. Details of formal training, such as lectures and short courses held at the universities and students who have worked at the Centre on Sandwich Courses or as vacation students, will be found in other sections of the report.

The termite group of the Biology Division donated live groups or small colonies of the North American damp-wood termite *Zootermopsis nevadensis* for class studies of the hind-gut protozoa to Imperial College, Kings College and Birkbeck College of the University of London, South Bank Polytechnic, North East London Polytechnic, the University of Aston in Birmingham and the University of Exeter.

East Anglia

There is continued liaison on the subject of African Armyworm with Dr D. J. Aidley of the Department of Biological Sciences and who is temporarily on an assignment at ICIPE.

Exeter

Dr J. M. Anderson of the Department of Biological Sciences visited the COPR Termite Group to discuss the role of soil-feeding species in decomposition processes. He was supplied with samples of preserved specimens for a pilot project to investigate the composition of their gut contents.

London

Birkbeck College

Continued cooperation by Dr E. Bernays with Dr Blaney of the College has resulted in extremely fruitful discussion and exchange of information on the sensory perception of locusts.

Imperial College

The ODM Liaison Officer in nematology, Dr J. Bridge, based at Silwood Park, provided valuable assistance with the nematology content of *PANS*.

Dr P. M. Symmons and Miss S. M. Green visited Silwood Park in April for discussions with Dr G. Conway and his colleagues about their work on mathematical modules of insect populations.

Dr N. Waloff of Imperial College agreed to act as one of two independent assessors for the work of the COPR Termite Group, which held a day-long seminar to describe its activities.

Dr Waloff is also Director of Studies for a COPR Research Fellow, Mr N. M. Collins, who worked throughout the year at Mokwa, Nigeria, on the energetics of *Macrotermes bellicosus*, a forestry pest.

London School of Hygiene and Tropical Medicine

Dr G. Webbe is a member of the COPR panel of consultants and has given valuable advice on the work of the Molluscicides Group.

Queen Elizabeth College

Mr R. W. Dunlop, a COPR Junior Research Fellow, is seconded to the Isotopes Unit of QEC to work on the synthesis of radioactively labelled molluscicides.

Queen Mary College

Dr J. M. Swift is cooperating with the COPR Termite Group through the Mokwa (Nigeria) research project to study the biology of the symbiotic fungus, *Termitomyces*.

Oxford

Dr W. A. Sands works with Dr J. Phillipson, Head of the Animal Ecology Research Group, to coordinate the termite research by Mr R. Buxton at Tsavo East Research Centre, Kenya with that in progress at ICIPE and in Nigeria at Mokwa.

Southampton

Mr C. Longhurst, a CASE Research Fellow funded jointly by SRC and COPR, has continued his studies under Dr P. Howse in the Department of Biology on the behavioural and ecological interactions between subterranean termites and their ant predators. Mr Longhurst works in cooperation with the Mokwa termite ecology scheme and is supervised and advised in his work by Dr Sands and Dr T. C. Wood. (See Section IVB.)

A new CASE Fellowship has been awarded to Mr P. H. Briner to study comparatively the defensive secretions of soldiers of Macrotermitinae.

Discussions have been held at the University's Unit of Insect Chemistry on the chemistry of leaf-cutting ant pheromones. Representatives of ICI, the Department of Zoology at Bangor and Rothamsted Experimental Station attended. Dr R. F. Chapman of COPR acted as Chairman and consultant at several meetings.

York

Dr W. A. Sands and Mr R. M. C. Williams of the Biology Division termite group have continued to exchange information and advice with Dr M. B. Usher, Department of Biology, University of York, formerly on contract to COPR's recent technical aid scheme in Ghana for the setting up of a unit at the Building & Roads Research Institute, Kumasi, to test materials against termite attack. Mr Williams and Dr Usher visited BRRI for two weeks in July to advise on testing projects.

Ten members of the Biology Department of York University have provided field assistance in the Crete armyworm project.

WEED RESEARCH ORGANIZATION

Mr D. R. Johnstone took part in an informal meeting to discuss present achievements and objectives in "Herbicide application in very low volumes", a project started under the supervision of Dr K. Holly at Kidlington in 1971. Mrs K. A. Johnstone attended the follow-up meeting in October to which technical representatives of agrochemical companies were also invited.

WRO also cooperates closely with the staff of *PANS* in ensuring an adequate supply of weed control information for the journal.

IV TECHNICAL REPORT

A LOCUSTS AND GRASSHOPPERS

1 ACRIDID TAXONOMY

Research on field populations of *Locusta migratoria migratorioides* (R. & F.) in Mali continued with work on an analysis of the results of a breakdown of field material into age groups post fledging and the characteristics of the male meiotic cycle. There is a preliminary indication that the number of supernumerary chromosomes increases in testis tissue during the period of adult dispersal just post fledging. This work was carried out by Dr N. D. Jago and Miss P. McAleer.

Mr A. Antoniou and Mr R. Dawson carried out reciprocal crosses of races of *Locusta migratoria* from Majorca (subsp. *cinerascens*) and Australia (subsp. unnamed). This was intended as an investigation into possible genetic isolation between races remote from each other in this species. Parallel cultures of the pure races were maintained as controls. Factors measured were:

(a) Incubation period of eggs
(b) Hatchling weight
(c) Colour patterns of all instars — later quantified
(d) Fledgling weight
(e) Adult morphometrics
(f) Chromosomes in male meiotic tissue in insects roughly 1 week post fledging
(g) Isoenzyme phenotypes — (esterase analysis of P. J. Curry)
(h) Adult fecundity and fertility

The F_1 generation was back crossed with parent stocks reciprocally as well as selfed. This enormous experiment has produced some intriguing results which will have to be analysed carefully. Chromosome preparations have still to be completed. No overt signs of genetical incompatibility were noted.

Data gleaned from esterase studies, chromosome studies and genetical crosses is essential as background before starting a taxonomic revision of the genus *Locusta*. The taxonomy section now maintains cultures of *Locusta* from S. France, Majorca, Saudi Arabia, Mali and Australia.

Considerable additions to the material needed for a faunal review of the Acridoidea of East Africa were made in August—September by Dr I. A. D. Robertson and his wife Mrs A. Robertson. COPR now has a fine, carefully annotated collection of Kenyan acridoids collected by them. Botanical and ecological notes accompany the collection. Dr N. D. Jago has started a revision of the genus *Usambilla* and genus *Kassongia* using the Robertson collection and his own material from Tanzania and Uganda. A number of such revisions will be necessary before a faunal work for Eastern Africa is published under joint authorship.

Taxonomic identification for COPR and the CIE were carried out in 1975 as part of routine in the section. Material originating from India, Nepal, Java, Sarawak, Malaysia, Brunei, Laos, Kenya, Nigeria, Cameroons, Rhodesia, Ethiopia, Malta, Iraq, Kuwait, Turkey, Saudi Arabia, Grenada W.I. and Canada was processed. A reference collection of Nigerian acridids was sent to the University of Ibadan. A small collection of gryllids from University of Dar es Salaam was identified and returned.

Revision of the genus *Oedaleus* (sub. f. Oedipodinae) was widened to include the genus *Gastrimargus*. The first is important as an economic genus including *Oe. senegalensis*. The second is primarily important because of its confusion with *Locusta*. Mr J. M. Ritchie, fellowship student i/c of these studies, has established a useful list of synonymy and examined most of the types in these two genera. Description and distribution maps are in preparation. Revision of the genus *Ochrilidia* (Gomphocerinae) was continued. Distribution maps and measurements were completed by Mrs S. E. J. Storer. Cabinet revision of the genera *Cataloipus* and *Heteracris* has been completed using mainly BM(NH) material.

Considerable re-organisation of the BM(NH) collections was taken in hand by Mrs S. E. J. Storer. Following the publication of a review of the sub. f. Oxyinae by Mr D. Hollis (formerly of this section) this group of genera was completely rearranged and rehoused (20 drawers). The Euryphyminae (12 drawers), Calliptaminae (54 drawers) and Pyrgomorphidae (101 drawers) were curated and relabelled, generic synonymy and grouping being based on revisionary works of Mason, Jago and Kevan respectively. The Calliptaminae were particularly difficult as *Calliptamus* species had to be organised with dissected genitalia rehoused in vials below the appropriate male insect. Mrs Storer also cooperated with Mrs J. MacDonald on the new catalogue of injurious grasshoppers.

Editorial and supervisory work reached a peak this year. Work by Mr P. J. Curry on the isoenzymes of field populations of *Locusta* in Mali has been prepared for publication. Dr N. D. Jago was involved in work for the second volume of '*Locusts & Grasshoppers*' by Sir B. P. Uvarov and a new catalogue of economic acridoids being directed by Mr J. Roffey. Dr Ali Soltani successfully submitted his thesis on the genus *Dociostaurus* (Gomphocerinae) and subsequently prepared a draft paper on the new species and taxonomic rearrangements for immediate submission to the Royal Entomological Society, London. Dr John Ohabuicke, counterpart research member on the OICMA staff in Mali, visited London in early September and Dr Jago then assisted with his thesis. The thesis was successfully submitted in December for the Ph.D. degree of Reading University. His subject, the seasonal nutrition of *Locusta migratoria* in Mali, is of great interest to our understanding of the population dynamics and hence control strategy for that species.

2 INSECT/HOST-PLANT RELATIONSHIPS

During this year the work on insect/host-plant interactions has continued. A major project on the food and basis of food selection of *Locusta* has been concluded and the work expanded into new areas. A study of the literature reveals that plants belonging to many different families may be eaten, but more critical analysis indicates that, outside the monocotyledons, plants are generally consumed only when the insect is suffering some kind of stress, such as lack of food or water. Within the monocotyledons, plants of several families are eaten, but, apart from the sedges and grasses, some degree of stress appears to be necessary before the plants are eaten whereas most sedges and grasses are readily accepted at all times. It is clear that *Locusta* is an oligophagous insect, feeding essentially on Graminae, but the work emphasises the necessity of careful and critical observation in field studies. At least in this species the degree of oligophagy is variable. Well-fed insects exhibit

Fig. 9. Solitarious locusts have to be reared individually. Here a Sandwich Course student is examining cultures in which they are in isolated Kilner jars.

a high degree of selectivity. With increasing deprivation of food or water more plants are eaten and the insect may become effectively polyphagous. It does not follow, however, even under extreme conditions, that these plants are readily accepted. It is known in some cases that less acceptable plants are eaten only after first being rejected, and even when eaten they may be taken only in small quantities, emphasising that the damage potential to unusual hosts depends on the numbers of insects to a very great extent.

Laboratory work in which over 200 plant species have been tested bears out the conclusions drawn for the field. Locusts deprived of food for only a short time reject almost all plants other than grasses, sedges and rushes; palms and some other monocotyledons are eaten, but only in small amounts.

The basis of selection has been shown to rest almost entirely on the presence of chemicals which inhibit feeding on non-host plants. These chemicals belong to a wide range of chemical classes and almost any one of them may prove inhibiting if it is present in high enough concentration. Most plants contain a range of inhibitors, the interactions of which remain to be studied. These conclusions emphasise the scope which is available in the development of crop plants which are resistant to insect attack. To date, most of our work has been concerned with factors which reduce or prevent feeding, antifeedants. But many plants which are eaten contain chemicals, alkaloids and steroids for instance, which might be expected to have metabolic, antibiotic, effects. This year preliminary studies have been made on the long term effects of selected chemicals. Dr Navon, a visitor from the Volcani Institute in Israel has been studying non-protein amino acids. These are known to have adverse metabolic effects on mammals and they commonly occur in leguminous plants, which, it has been noticed, often lead to moulting failures in locusts if they form a large proportion of the diet.

3 REPRODUCTION AND DEVELOPMENT

Dr A. R. McCaffery has concluded his studies on the control of oocyte development in *Locusta migratoria*. Previous work had shown that fully primed corpora allata and an adequate diet are pre-requisites for successful egg development.

The results of experiments carried out this year have slightly modified initial ideas on the role of the cerebral neurosecretory system during vitellogenesis. Removal of the median neurosecretory cells prevents yolk deposition entirely but does not prevent normal pre-vitellogenic oocyte growth. Implantation of active corpora allata into females with fully electrocoagulated median neurosecretory cells does not lead to the production of mature oocytes but does allow vitellogenesis in up to seventy percent of females. These vitellogenic oocytes begin to resorb when half-way through their development and oviposition does not take place.

It has been concluded that neurosecretions from the median cells of the pars inter-cerebralis play no direct physiological role in vitellogenesis. By activating and maintaining the corpora allata, the median neurosecretory system has an indirect influence on vitellogenesis. Thus in females in which the neurosecretory cells have been removed the corpora allata remain inactive and egg development is inhibited. Vitellogenesis occurs when already primed corpora allata are implanted into these females. Since stimulation of the glands cannot be maintained, the implanted corpora allata become inactive and oocyte resorption occurs. The failure of some females to respond to such implantation treatment is probably due to a number of side effects associated with electrocoagulation of the median neurosecretory cells. This technique has been improved and mortality has been reduced. It is likely, however, that nervous pathways associated with feeding can be affected by the electrocoagulation. To demonstrate the effect of lack of food, neck-ligatured females were implanted with active corpora allata. Following this treatment there is a very slight amount of oocyte growth and yolk deposition in a proportion of females. Lack of food has very obvious inhibitory effects.

A series of experiments has been undertaken in an attempt to determine the nature of the control of the cyclic development of successive oocytes in an ovariole of *Locusta*. From simple observation of the cycle of egg development it is clear that growth of sub-terminal oocytes begins well before ovulation of the terminal oocytes. Furthermore, a yolk-like material is laid down in these sub-terminal oocytes prior to terminal oocyte release. It is thought that this material is derived from haemolymph yolk proteins liberated during resorption of a proportion of terminal oocytes.

By ligaturing the lateral oviducts at different points it has been possible to prevent oviposition of mature eggs. These eggs are retained in the oviducts. Subsequent development of further batches of eggs continues normally and these too are released into and retained by the lateral oviducts. Egg development continues in this fashion. The lateral oviducts may finally split, spilling normal mature eggs into the body cavity. It appears that the presence of mature oocytes in the oviduct has little or no effect on the development of subsequent oocytes.

When the base of an ovariole is ligatured so that release of single mature oocytes is prevented the sub-terminal oocyte begins development normally. Resorption of the retained oocyte often occurs. The presence of the retained first oocyte has little or no effect on subsequent oocyte development although resorption of both the terminal and the immediately adjacent sub-terminal oocytes often occurs.

Ovarioles cut from the lateral oviducts continue development normally in the body cavity although their development is often slow. It is apparent that the ovarioles do not need a direct connection to the oviduct for normal development to occur.

At present it seems that the cyclicity of oocyte development is not governed by the ovary or the integrated structure of the ovary but by the inherent developmental stage of the oocytes themselves. The largest and most competent oocytes are at the base of the ovarioles and, under hormonal influences, they become vitellogenic. In conditions where excess proteins or hormones are available the next oocytes begin development.

Dr D. W. Ewer, in extending his work on the water relations of egg pods, has been re-examining the problems of the function of the specialised area at the posterior pole of locust eggs. According to some previous workers both water entry and water loss occur only from this region of the egg; it has therefore been called the hydropyle. Other experimenters have failed to find evidence for any difference in water permeability between this region and the rest of the egg surface and have concluded that the hydropyle is not concerned with water uptake. In all these investigations the eggs have been subjected to treatments which might have secondary effects upon their physiology and it appeared desirable to examine the question using procedures which did not involve special experimental treatments of the eggs.

The problem of whether there is a polarization of the egg in relation to water uptake has been approached by studying weight changes in eggs with either only the anterior or only the posterior pole of the egg in contact with damp sand. Eggs do not start to take up water from the soil until after about 75 hours development at 30°C. If eggs of *Schistocerca* which have developed for about 60 hours are separated from an egg mass and orientated so that the hydropylar region is in contact with damp sand, water uptake follows a normal pattern. If the eggs are orientated so that only the anterior pole is in contact with the sand, no water uptake occurs and development ceases before blastokinesis. If such eggs are then reversed so that the posterior pole is in contact with the sand, water will be taken up and development proceeds. It thus appears that the hydropylar area of the egg is concerned with water uptake, at least during the first stage of this event. It should, however, be realised that this could be due to the special structure of the chorion at the posterior pole: it does not show that the posterior cuticle has special permeability properties.

Such results are obtained only if the eggs are kept in a sub-saturated atmosphere. In 100% relative humidity, water loading occurs regardless of whether the eggs are orientated with the anterior or the posterior pole in contact with the sand. This difference may be due to the fact that the chorion of the egg can act as a wick conducting water to the posterior pole. In a sub-saturated atmosphere water evaporates from the chorion too fast to allow it to move from the anterior to the posterior pole.

Similar experiments using eggs which have been incubated for about 120 hours before separation give a different result. Even in a sub-saturated atmosphere such eggs are able to take up water when the anterior pole is in contact with damp sand. This suggests that there has been a change in the permeability of the egg surface to water. Later still, once the eggs have taken up water for a period of about 60 hours, permeability falls again.

These results, which are in agreement with other observations which suggest that the hydropyle is concerned with water uptake, imply that if *Schistocerca* oviposits in relatively dry soil, the eggs remain relatively impermeable to water and enter a phase of developmental quiesence. If water becomes available, it first enters the egg through the hydropyle. If the supply of water is adequate, the whole surface of the egg becomes transiently permeable to water. Water loading is thus initiated by water supply and only if this is adequate does the egg pass to the potentially dangerous stage of high permeability needed for rapid water uptake.

Mr D. J. Chamberlain has been continuing his studies on environmental influences affecting the viability of locust eggs in their pods. Close crowding of pods can occur in the field. Earlier experiments showed that this lowers the viability of the eggs of *Schistocerca* as a result of soil compression leading to rupture of individual eggs. Mr Chamberlain now finds that the foam surrounding the egg pods of some species of locust acts as a 'buffer' which prevents sand particles from rupturing the eggs. In extreme conditions of crowded oviposition, the naked eggs of *Schistocerca* have no protection against such damage and viability is low; the viability of the foam-covered eggs of *Locusta* and *Locustana* is greater, the dense foam covering around the eggs of *Locustana* giving better protection than the slightly less rigid covering around the egg mass of *Locusta*.

Ruptured eggs serve as a focus of infection which spreads through a pod from the point of rupture. If this infection is established during the early stages of development it has a greater effect upon adjacent eggs than if the infection occurs during later embryonic development. Fungi may grow at such sites of infection. They do not, however, affect intact eggs and are probably feeding on nutrients available

Fig. 10.

Egg pods of *Schistocera gregaria*. Left normal egg pod laid in damp sand; right, pod laid in 9 cm depth of 8–12 mesh sand (10% water v/v) overlying sand of less than 60 mesh (15% water v/v).

Fig. 11.

Dr. A. Navon, a visiting research worker from Israel, who was with the Centre for one year, is studying the results of a feeding experiment with locusts.

from ruptured eggs; hatching from eggs overgrown with fungal hyphae is usually successful. It thus seems probable that bacteria or perhaps virus particles, are responsible for the infection which spreads from ruptured to healthy eggs.

In experiments with crowded egg pods of *Schistocerca*, it was noted that a considerable proportion of the hatchlings which emerged were green and remained green, rather than turning black. If crowding of the pods occurred about four days after laying, nearly 50% of the hatchlings were green and small and none survived beyond the second nymphal instar. Examination showed that the eggs on the outside of these pods were scarred and that water had been lost from these eggs. In experiments in which the water content of partially water-loaded eggs was deliberately reduced by exposure to dry air, the hatchlings were on average 10 mg lighter than the controls and they failed to turn black; the vast majority of such hatchlings died within 24 hours of eclosion. Thus even if there is no rupture of eggs as a consequence of crowded oviposition, a high percentage of the hatchlings may have reduced viability. In *Schistocerca* at least, there are two distinct mechanisms which tend to reduce the potential productivity of egg fields in which dense oviposition has occurred.

Both soil types and soil gradients affect egg pod morphology and the subsequent survival of first instar nymphs. *Schistocerca* will not oviposit in very coarse gravel and shows great reluctance to lay in wet, very fine soils. If the females are offered soil in which the upper layer is acceptable but the lower layer is not, they will oviposit. The resulting pods are, however, stunted, with a short foam plug and a greater number of eggs per centimetre length of egg mass than is normal (see photo). The eggs in these stunted pods rarely hatch as they are close to the soil surface and almost invariably die from water loss.

The presence of egg pods of both *Schistocerca* and *Locusta* affect the ability of water to move through the soil. If the soil around a pod is allowed to dry out and water is then made available from below, the water will rise through the soil. This movement occurs much more rapidly in the soil immediately around an egg pod. Since it has not been found possible to imitate this effect by mechanical compression of the soil, it is provisionally concluded that some hydrophilic substance, secreted during oviposition, spreads into the surrounding soil.

Mr Padgham has now completed his work on the physiology of melanisation of first instar *Schistocerca* nymphs. By electrocoagulatory destruction of the nervous system posterior to the metathoracic ganglion it is possible to produce first instar nymphs which do not melanise. Additional evidence that this is the location of the hormone was obtained by making high-potassium Ringer extracts of different parts of the nervous system. High-potassium Ringer induces the release of neurosecretion and it was found that only extracts from the area of the metathoracic ganglion caused melanisation. It has been further shown that the hormone is analogous to bursicon, the tanning hormone. Moulting larvae can be ligatured so as to limit melanisation to only one side of the ligature. Histological examination of such larvae showed that the distribution of tanning was the same as the distribution of melanisation. Both melanisation and cuticular sclerotization show identical activity curves, both are initiated by a blood-borne factor released at the time of moulting and both show a progressive decline in activity over the following 3 to 4 hours. The standard blowfly bioassay for bursicon was used to assay the blood of first instar nymphs; this was positive and from all this evidence it has been concluded that bursicon, released from the nervous system immediately posterior to the metathoracic ganglion at the time of the intermediate moult, controls both melanisation and sclerotization.

Having elucidated the control process in normal gregarious larvae, some non-melanising forms were examined. In addition to albino locusts there are often a few non-melanising larvae in a normal gregarious pod. Both of these non-melanising types were examined and in each case blood samples were found to be capable of inducing melanisation in normal vermiform larvae. When, however, either of the non-melanising forms was injected with blood from normal gregarious larvae they did not melanise. It is concluded that the failure of the non-melanising forms to melanise is due to an inability of the epidermis to respond to the hormone rather than to an absence of the melanising hormone in the blood.

Work now in progress is a study of the control of egg pod hatching using both field and laboratory data. The field data were collected in Saudi Arabia and confirmed the suggestion of other field workers that hatching occurred at dawn. In the laboratory it has been found that egg pods exposed to a diurnal temperature cycle similar to that experienced in the field can be induced to hatch some hours earlier than normal by prematurely lowering the temperature during the last hours of incubation. The inference from this is that a drop in temperature is the stimulus for hatching. Additionally, if egg pods are exposed to a diurnal temperature cycle until the last day of incubation and then kept at a constant high temperature, the eggs still hatch at the normal time. There thus seems to be entrainment of the hatching process by the temperature cycle.

In the field the eggs are conditioned to hatch at dawn by experience of the diurnal temperature cycle; this is reinforced, and can to an extent be overridden by, the cold stimulus on the last day.

Future work will include a study of other responses to variations in the temperature cycle and also the physiological control of this process.

4 NEUROPHYSIOLOGY AND FLIGHT

The studies on visually mediated flight behaviour have continued. An attempt has been made to characterise the properties of the stimulus which are most effective in eliciting the yaw response discovered last year. It has been shown that with gross stimulation the yaw response is replaced by an escape response. The insect stops flying momentarily thus losing height and moving from its previous flight path. This may be adaptive under field conditions enabling the insect to escape from predators whilst in flight. To define more closely the insect's ability to maintain stability in the yawing plane apparatus was assembled to enable the experimenter to rotate the flying insect across the laminar air flow in the wind tunnel. For simplicity of analysis the insect was rotated sinusoidally about the centre line of the tunnel, the angle and rate of displacement being controlled over a wide range of values and rates. Preliminary analysis shows that muscle firing is synchronised with the movement of the whole insect such that, in the second basalar muscles at least, firing on one side stops and restarts in the corresponding muscle on the opposite side as the insect approaches the centre line of the tunnel. This firing is then maintained, throughout the traverse and back to just before crossing the centre line again whence the change-over is repeated. Head rotation and abdomen movement occurs as does unilateral hind leg lowering, this latter response is so marked in some animals that the hind legs can be induced to retract and extend alternatively in a continuous rhythm at the same frequency as the forced yawing. The muscle responses occur in total darkness and must therefore be mediated by the wind sensitive hairs on the head of the insect. Removal of the antennae does not affect this behaviour but covering the head with wax does disrupt the response. The stimulation of these responses and their control is being studied.

Together with these neurophysiological and behavioural studies high speed cinephotography has been used this year and has enabled us to see how the wings move during the visually mediated yaw response. The response is signalled at the top of an upstroke, the wings remain in this position for one cycle before descending on the downstroke. In the next cycle, the downstroke is extended more deeply and the wing is pushed further forward on that side of the body away from the stimulus. It is at about this time that the yaw movement can be recorded, subsequent wing strokes are less obviously enlarged and the wing beat cycle returns to a pattern similar to that existing before the stimulus in about five further cycles. Thus the response is completed in under one quarter of a second. The temporal relationship between stimulus reception and response occurring is being studied and aided by recordings from the central nervous system during flight.

It is planned to extend the observations of the flight behaviour of locusts into the field (Australia) in early 1976 where it is hoped to film swarms of *Locusta* using the high speed cinephotography techniques developed this year.

5 DESERT LOCUST RESEARCH

The Desert Locust Forecasting Project

Following the resignation of Dr L. Bennett, Mr Roffey was appointed to lead work on the Project which is aimed at producing a manual on the principles and practice of forecasting changes in the distribution and size of desert locust populations for use by the regional and national forecasting offices.

Preliminary work on recession situations in the Eastern Region by Mrs Steedman, Mr Douthwaite, Mr Lambert and Mr Bettany were halted pending a comprehensive review of information already available and what further information was required for the manual. This review was nearly complete at the year's end.

The use of satellite imagery to detect potential desert locust breeding sites

A major problem encountered by all organisations concerned with surveying and controlling desert locusts is the absence of rapid, reliable and comprehensive information about where rain, which is necessary for multiplication, has fallen. At the request of Locust Control Office of FAO Mr J. Roffey prepared a programme plan for improving Desert Locust survey and control using satellite application techniques, which was duly accepted by the 19th Session of the Desert Locust Control Committee in October. The object of the programme is to develop a co-operative international project in which satellite imagery, interpreted in the light of ground truth data from restricted areas, can be used to provide rapid, routine, comprehensive, accurate and inexpensive information on the occurrence of potential desert locust breeding sites.

6 ZONOCERUS VARIEGATUS IN NIGERIA

The 1974—75 season was one of particular interest in the *Zonocerus* research. On the basis of the previous season a five-fold increase in the population had been predicted and the prediction proved correct. As a result of the enormous population, defoliation of the cassava occurred very early in the season and extensive loss of crop seemed likely. However, an unusually early start to the rainy season, at the beginning of February, led to an outbreak of the fungus *Entomophthora grylli* which reduced the population by a half in 10 days. This relieved the pressure on the cassava, which at the same time regenerated strongly in the wet conditions so that total loss of yield was probably slight. This occurrence highlights the unpredictability of the importance of *Zonocerus* from season to season, as have events in the 1975—76 season.

Apart from the general population studies, which were continued in 1974—75, three main aspects of the biology of *Zonocerus* were investigated:

1. The effects on yield of artificial defoliation of the cassava crop.
2. Reproductive potential.
3. Causes of natural mortality.

Yield was studied in three areas in which the plants were artificially damaged. Three treatments were applied in addition to the control:

1. A single defoliation to correspond with defoliation in the field.
2. Defoliation at 2-weekly intervals or six weeks such as might be caused by successive waves of insects passing through the plots.
3. Cutting off the stem above the second notch — simulating the extreme damage caused by high populations.

The first treatment produced no significant reduction in yield, but both the others did so, in some cases by as much as 50%. Parallel studies of natural damage to cassava indicate that these levels of damage are caused by *Zonocerus* in some situations. There is no doubt that under these circumstances *Zonocerus* is very important indeed. The unknown at present is the frequency with which these circumstances occur. Studies in the 1975—76 season should provide information on this aspect.

The reproductive potential of *Zonocerus* was studied in the field and in the laboratory on different foods. The laboratory studies showed, as have previous studies on feeding, that although *Zonocerus* eats a wide range of plants, survival and development is generally poor except on a few species, of which cassava is one. Even on this plant the production of eggs falls far short of the potential number possible if all the ovarioles produced viable oocytes. Many oocytes fail to develop at all and others are resorbed. Similar levels of egg production were found in the field where few of the insects laid more than one egg pod. The eggs laid by long-winged females hardly contributed at all to the population.

Two main causes of natural mortality were studied, the fungus already referred to and the parasitic fly *Blaesoxipha filipjevi*. The fungus affects large numbers of insects about 8 days after heavy rain, the insects dying in the late afternoon clinging to the vegetation of the roosting sites. Here they are easy to count and act as foci of infestation for other insects. At peak periods 10—15% of the population may die in one day with other peaks following at approximately 8 day intervals — the incubation period of the fungus. With continued periodic rain the outbreak persists, but in the absence of rain the infection slowly dies away. Parasitism by the fly only begins in the adult stage, and does not reach high levels until the insects aggregate for oviposition. At this time the adult flies are seen sitting amongst the grasshoppers which provide easy targets. The level of parasitism builds up rapidly until as many as 70% of the population may be affected. This parasite causes the final decline of the population and commonly prevents more than a single oviposition by each female grasshopper. It has relatively little effect on the number of insects laying one pod because it does not become common until after this.

Other aspects of the biology which were studied were the development of oviposition groups, sound production by males and the occurrence of flight by the long-winged insects.

Experiments in control of the later instars by normal insecticidal methods were also carried out. As with the early instars, BHC and fenitrothion were shown to be highly effective and populations were readily controlled in field trials. However, the mobility of the insects was so high that four or five sprays were needed in a month to keep the population at a low level. Parallel studies on movement using the radiotracer P^{32} indicated a 50% turnover of the population of cassava plots is about five days. The insect is more mobile than simple counts may suggest and this makes economic control by insecticides during the later stages difficult.

The aggregation of the population for oviposition presents another potential source of control. In two areas, each of approximately 1 km^2, oviposition sites were located by surveys during the oviposition period. Subsequently the egg-pods were dug up, some hundreds of thousands being destroyed in each locality. A second survey after hatching indicated that in both areas over 90% of the population had been destroyed. The ultimate effect of this work on damage during the next season remains to be assessed, but the method has potential as one which could be employed by the farmers themselves without special techniques. The main oviposition groups are easily recognised and occur at a time when the farmer is actively cultivating and so spends more time in the fields than usual. If the current season's results prove satisfactory an attempt will be made to co-operate with farmers and to test their ability to find the eggs.

During the year Mr Page was in permanent residence in Nigeria. Visits of varying duration were paid by Dr Bernays, Dr Chapman, Dr Harris, Dr Hunter-Jones and Mr A. Mitchell. The group actively co-operated with Dr Yondeorei in the Department of Agricultural Biology and Dr Moddu in the Zoology Department of the University of Ibadan, and discussed this work with other scientists in Nigeria working on *Zonocerus*.

7 LOCUSTS AND GRASSHOPPERS OF ECONOMIC IMPORTANCE IN THAILAND

A paper summarizing information, much of it unpublished, concerning the 55 species of locusts and grasshoppers of economic importance which occur in Thailand was completed during 1975. Most species are of very minor importance but with the rapid clearing of forest several have become locally important pests of upland rainfed crops. The most important species economically is the Bombay Locust, *Patanga succincta* L., which has become a major pest of maize, an important export commodity. There is no evidence that species have become economically important due to immigration from outside Thailand, although both the Bombay Locust and the Oriental Migratory Locust, *Locusta migratoria manilensis* Meyen, are migratory elsewhere in south and south-east Asia.

To enable the Locust Control-Research Centre, Thailand to improve their control strategy some aspects of the distribution of *Patanga* were studied. Mr R. Steedman with the help of Mr Wit Namruangsri collected records of occurrence of the insect in Thailand and produced monthly maps of the distribution of adults and of hoppers. A study of a particularly well documented control campaign in a small area was made and the results tentatively suggested that the beginning of laying could be related to particular falls of rain. Mr R. A. Steedman carried out trials on the rate of reinvasion of sprayed plots and found that within 10 days of spraying six one-hectare plots with fenitrothion 83% ULV numbers of adult locusts were as great or greater than before spraying. Mr R. J. Douthwaite studied the diurnal behaviour of the nymphs. Greatest densities occurred where the low-growing weed *Brachiaria reptans* grew amongst taller maize. The nymphs sheltered in the *Brachiaria*, the younger ones eating *Brachiaria*, the older ones, maize. More effective weeding might therefore reduce locust numbers. The number of nymphs on the maize plants increased with temperature over the range $24°-33°C$. and normally therefore greatest numbers were exposed to airborne insecticide during the mid-afternoon.

B TERMITES

1 BIOLOGY AND TESTING

The extraction of living cultures of the dry-wood termite building pest genus *Cryptotermes* from infested pine and other lumber, imported from West Africa at the end of 1973 by Mr R. M. C. Williams, was completed by Mr M. J. Pearce in 1975.

This *Cryptotermes*-infested pine wood still contained flourishing colonies in the resinous heartwood, most of the sapwood having been eaten. An experiment was begun to test the ability of *Cryptotermes* to colonise heartwood in the absence of sapwood. The early results suggest that heartwood is not readily colonised.

The tests of *Cryptotermes* colonisation ability on 8 timber species reached a total of 9 replications each for *C. brevis* and *C. havilandi*, with the addition of 88 individual colony replicates. The results of earlier replications show clearly that *C. brevis* has been unable to colonise Obeche (*Triplochiton scleroxylon*) or Ramin (*Gonystylus bancanus* and spp.), although these could be fed on by dealate pairs and are readily eaten by large colonies when other woods are also available. Thus these timbers are not repellent, but appear nutritionally deficient in some way. Obeche seems to become satisfactory after fungal degradation. The experiments with *C. havilandi* are less advanced but are giving similar results. This species suffers heavy early mortality among colonising pairs, in contrast to the low production and elimination rates reported last year for neotenic reproductions, and again differing markedly from comparatively successful colony foundation rates of *C. brevis*. This may be another biological difference between them related to climatic preferences which are also being studied.

Climatic stress limits for the three African *Cryptotermes* species have been worked out by Mr Williams from their known distributions, and the corresponding published meteorological data. A new method for calculating extrapolations from the standard daily readings had to be developed to provide the required predictive parameters. Recent new records from two new localities have confirmed their value. Experimental work has been designed to check and refine the concept in the laboratory. Humidity tolerance tests of colonisation ability have begun, using ten imago pairs per replicate in wood blocks kept at $28°C$ in ten relative humidity levels. These span the range of annual mean saturation deficits and wood moisture contents found in building timbers in all parts of Africa where *Cryptotermes* occur.

The studies of alate production as a response to diminishing food resources have continued and although some of the *C. brevis* colonies have been constantly monitored throughout 1975, no alates have yet appeared. The first four replicates of a parallel test with *C. havilandi* have been set up. Owing to a shortage of alates in this species small groups of pseudoworkers have had to be used.

The long-term development of colonies is being studied comparatively in *Neotermes jouteli*, a dry-wood termite of tree crops and forestry in the Caribbean, and the two species of *Cryptotermes*. One hundred colonies of *C. brevis* and twenty of *C. havilandi* have been set up in this part of the project.

The emphasis in the biological programme is placed throughout upon the production, behaviour, and survival of colonising alate reproductives, and the subsequent colony development. This is fundamental to understanding how infestations in furniture and building timbers grow and spread to economically-damaging levels. It represents an approach to these pests not previously tried, owing to the difficulty in sustaining adequate supplies of fresh alate reproductives. This depends upon the maintenance of large numbers of flourishing colonies that can regularly convert a major proportion of their standing crop biomass into the dispersive phase. The total number of live cultures of all species rose from 248 to 585, most of the increase being accounted for by 222 colonies of *C. brevis* and 146 of *C. havilandi*. Dr Sands brought from Sri Lanka a large culture of *Postelectrotermes militaris*, an important pest of growing tea. This is of great interest because of its high tolerance of supernumerary neotenics, and the original consignment has quickly produced 6 viable subcultures. The species not involved in current research are only kept in small quantities to provide for donations to other cooperating institutions or for teaching purposes. Universities and Polytechnics depend upon COPR for course-work material, and demand for *Zootermopsis nevadensis* was as high as ever. Major donations of live breeding colonies for research were also made to the following: Dr J. M. Anderson, University of Exeter, *Z. nevadensis;* Dr F. L. Carter, USDA Southern Forest Experimental Station, Gulfport, USA, *Reticulitermes flavipes santonensis;* Prof. Dr M. Luscher, Abt. fur Zoophysiology, Bern. University, Switzerland, *N. jouteli;* Prof. Dr G. Becker, Bundesanstalt für Materialprüfung, Berlin, West Germany, *C. havilandi*.

Materials testing for British exporting industrial concerns has continued as a minor aspect of the work with live termites. Plastics used in electric cable sheaths have apparently stood up well to the 'stress' techniques described in an earlier report. Some soft plastics remained undamaged, leading to a reappraisal of the method. It appears that termites might be able to detect with their mouthparts elastic stresses in materials, a speculation that requires testing by further research. The performance of similar cable samples exposed in the main culture tanks has yet to be assessed. A test of polyurethane foam insulating mouldings was completed. The final bioassay of the moulting inhibitors, PH 60–40 and PH 60–38 for Philips Duphar Limited using both compounds at 300 ppm against *Zootermopsis nevadensis* and *Reticulitermes lucifugus* demonstrated a weak effect by PH 60–38 against *Z. nevadensis* only. With both compounds tested at the same strength against *Cryptotermes brevis*, only PH 60–40 produced a weak effect.

2 TAXONOMY AND MORPHOLOGY

The steady increase in the biological and ecological activities of the group has had to be partly at the expense of progress in taxonomic and morphological research. Revisionary work on the genera *Microtermes* and *Amitermes*, and the functional morphology of mandibles, remain the chief research interests of Dr Sands. The revision of *Bifiditermes* by Mr Williams and his West African keys, have had to be shelved for the moment.

Mr S. Bacchus has continued his work on the functional morphology of termite legs.

3 IDENTIFICATION AND ADVISORY SERVICES

The identification service continued at a turnover rate similar to the previous two years. It remains the primary responsibility of Mr Bacchus, with occasional help from other members of the group. The collections were on the whole larger, from a total of 15 developing countries.

The time-consuming technical and scientific advisory service is run by Dr Sands and Mr Williams jointly and is quantified for the first time. The turnover in 1975 was 102 enquiries, virtually all of these originated in 21 developing countries, and they ranged from the susceptibility of materials to termite attack, to requests for advice on research programmes. Particularly noteworthy were the 12 enquiries from British manufacturers, building contractors, or architects associated with export projects to OPEC countries of the Middle East. The most spectacular of these was a request for advice on termite-proofing the timber supported domed roof of the main assembly hall of a new Parliament building in Riyadh, Saudi Arabia. The competitive advantage to British industry in having a ready source of such information can hardly be over-stressed.

4 FIELD RESEARCH IN NIGERIA

The research into termite ecology at the Mokwa Agricultural Research station of Ahmadu Bello University, Nigeria, continued throughout 1975. The primary objective is to quantify subterranean termite populations, relating these to soil fertility and crop damage by the pest species. New approaches to control measures are also being studied to reduce the risks of environmental damage and improve their economic viability.

The research scheme works on a cooperative basis between ODM (COPR) and the Institute for Agricultural Research (ABU), Zaria. The team was led by Dr T. G. Wood until his return to the UK in June. He joined COPR permanent staff in October and continues to supervise the scheme, Dr R. A. Johnson having succeeded him as Team Leader. Mr C. Ohiagu is employed by IAR as Nigerian counterpart entomologist to the Scheme.

The sampling by Dr Wood of subterranean termite populations on long-term experimental plots set up last year has begun to give a clearer picture of the relationship between crop damage and population density. The principal problem is damage to maize by *Microtermes* which also attacks groundnut and sweet potato; yams are damaged by *Amitermes*. Four levels of attack on maize were categorised as follows: (1) damage to root system only, (2) damage to stem and roots, (3) plant half-lodged and (4) plant totally lodged (ie flat on the ground). The last of these is the only category contributing to serious loss in yield. Populations of *Microtermes* have risen from $300-1000/m^2$ in the 1st year, their attacks on maize being associated with 0.4–1.2% lodging, to $1320-2550/m^2$ in the 2nd year, with 1.1–3.1% lodging. In fields cultivated for 9 years populations of $2913-3939/m^2$ caused 29–33% lodging with yield loss of 5–8%. In 25 year old cultivation populations reached a maximum of $5652/m^2$, lodging 14% and yield loss 10%.

The bait sampling method developed by Dr Johnson has been used alongside the core sampling. Population estimates from the latter correlate closely with percentages of baits attacked.

The same sampling techniques were employed to provide population estimates of termites on the plots at IITA Ibadan belonging to the COPR Pesticides Residues Project. These were too small to be subjected to the disturbance of coring and baits had to be used. Cores were taken for comparative purposes from larger adjacent plots, and bush sites were similarly sampled. A bush population of $3163/m^2$ (all species) from cores was associated with a 30.9% attack on bait blocks. Population

Fig. 12. *Macrotermes subhyalinus* mound opened for population sampling, Kajiado, Kenya, ICIPE Termite ecology project.

Fig. 13.
Fungal spores (*Termitomyces* sp.) taken from the crop of the female reproductive of *Microtermes usambaricus*. The spores survive in the gut for several weeks and are used to inoculate the new fungus comb.

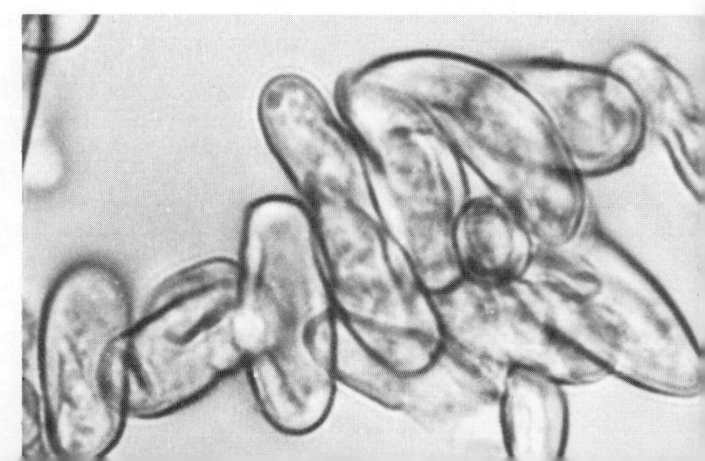

estimates from untreated cultivated and bush plots adjacent to the IITA experiments were less reliable owing to restricted replication, ranging from 667/m^2 to 5938/m^2, associated with 11.0–46.0% and 11.0–25.8% attacks on baits respectively. These compare with 0.0-3.6% in contaminated (DDT) plots and zero on treated plots. This evidence that crop spraying also profoundly affects termite activities indicates the need for further study and the importance of the research at Mokwa.

Dr Johnson has continued to study foraging activity in the subterranean Macrotermitinae by baiting techniques. Softwood baits had to be substituted for the ceiling boards which proved costly and irregular in supply. *Microtermes* have been found to show marked seasonal changes, being active near the surface only when soils are moist. The search for food is initiated by minor worker castes, which quickly recruit major workers when it is found. Both castes gather food, but major workers are much more efficient, and comprise 80% of foraging parties, compared with 15% minor workers and 5% soldiers.

Fungus combs of *Microtermes* sampled in pits of 2m^2 x 1 m depth on cultivated land have shown seasonal fluctuations in vertical distribution and biomass. The dry weight of fungus combs at the end of the wet season was over 10 g/m^2, of which more than half was in the top 50 cm of soil, compared with less than 0.1 g/m^2 at the end of the dry season. Since maize crop residues are the only available food, their removal should reduce termite populations; this is being tested experimentally.

At Shika Research Station, near Zaria, Dr Johnson began a study of termite attack on the pasture improvement legume, *Stylosanthes gracilis*. Sampling pits gave *Microtermes* fungus comb dry weight of 27.2 g/m^2, nearly 3 times the maximum Mokwa figure. Populations from soil cores were not proportionately high. This genus apparently has a highly adaptable symbiotic economy that may contribute to its ability to survive and achieve pest status in a wide range of conditions.

The laboratory rearing of Macrotermitinae has revealed the colony-founding females (not males) of *Microtermes* carry an inoculum of spores of their Basidiomycete fungal symbiont *Termitomyces*. A tightly-packed bolus of conidia survives a slow passage through the gut, to be deposited on the primordial comb produced by the first worker castes to develop. Studies of several other genera have not demonstrated such inocula, except in *Macrotermes bellicosus* where it is apparently carried by males alone. The related *M. subhyalinus* lacks the inoculum. These highly significant preliminary findings require confirmation.

The laboratory rearing is now reliable enough to consider its use for certain insecticidal bioassays. Preliminary bioassays of a range of fifteen insecticides were made by Mr A. L. Davies who also trained an assistant employed by ABU, during two visits to Mokwa. Topical applications failed because termites were too sensitive to both handling injuries and the toxicity of the commonly used solvent, AROMASOL H. Acetone though non-toxic, proved too volatile for accurate dosage measurements. A "dry-film" technique in a first screening test showed *Cubitermes* (a soil feeder) to be much more sensitive to both pirimiphos ethyl and DDT than *Trinervitermes* (a grass harvester). Pirimiphos methyl was less toxic to *Cubitermes* but far more toxic to *Trinervitermes*.

Microtermes in a comparable experiment was intermediate in sensitivity to pirimiphos methyl, much less sensitive to pirimiphos ethyl, and similar to *Trinervitermes* in its reaction to DDT and dieldrin. *Microtermes* alone was used to screen the entire range of insecticides, toxicity being assessed in terms of the strength of surface deposit required to produce 100% mortality after 24 hours exposure.

Fig. 14.
Macrotermes bellicosus, angry major soldiers on part of mound wall. The defensive secretions of this and related species are being studied by a CASE research fellow partly funded by COPR in connection with the Mokwa termite ecology project.

Fig. 15. Ethiopia. Dr. W. A. Sands with members of the Department of Argriculture conducting a termite survey of farmers' fields.

Dieldrin, heptachlor, and Dursban (chlorpyriphos) were far more toxic by direct contact than any other chemicals tested. Pirimiphos methyl, Carbofuran, propoxur, and bendiocarb were in a middle range, while DDT, Mirex, chlormephos, pirimiphos ethyl, chlorfenvinphos, and Leptophos were, in that order, decreasingly toxic. These results require confirmation and the insecticides will also have to be presented in other ways such as spiked baits and soil mixtures before making a selection for field trials on a larger scale.

The field studies of *Trinervitermes geminatus* as primary consumer-competitors of cattle by Mr C. Ohiagu were completed and are being written up. Populations of $210/m^2$ corresponded to a mound density of 232/ha, of which 100/ha died in a year and 64 were lost through cattle trampling and erosion. In the same period 44 new mounds/ha developed. Turnover was lower in ungrazed plots. Foraging by the termites was restricted to 182 days per year with a daily duration of 2–3 hours. Grass and litter production were 3157 kg/ha and 1406 kg/ha respectively of which cattle consumed 1404 kg/ha and *Trinervitermes*, 81 kg/ha. Laboratory tests of feeding preferences with transplanted nests showed a strong preference for *Andropogon gayanus* which is also preferred by cattle.

The work at Mokwa on *Macrotermes bellicosus* energetics by Mr N. M. Collins is nearing completion. Population estimates from 14 mounds ranging up to 5.8 metres high and 2.5 m hive diameter have provided a regression from which population density can be calculated. Over 50% turnover of the 6–7 mounds per hectare by death and replacement was recorded in 1975. Confirmation by the 1976 count would fundamentally change our view of the ecology of this species. Fungus comb turnover appears complete every 6 months.

Wood- and leaf-litter fall have been measured at approximately 1400 kg/ha each. Termites removed about 50% and 4% of these respectively, *M. bellicosus* being responsible for 21% and 3.4%. Oviposition rates have been estimated, and the respiration of both termites and fungus comb measured.

A study of behavioural and ecological interactions between termites and their ant predators by Mr C. Longhurst continued, with two visits to Mokwa interspersed with work at Southampton University. Predation rates by the ponerine raider *Megaponera foetens* that specialises in attacking foraging parties of Macrotermitinae have been estimated. Efficiency of predation varies with the termite species attacked which change seasonally. It remains sufficient to turn over an important proportion of the termite standing crop biomass in the course of the year. Trail following by the raiding ants has been examined, and it appears that dispersing male reproductions also use the workers' foraging trails to locate nests, which they enter, presumably to fertilise existing or virgin queens. The transplantation of *Macrotermes bellicosus* colonies to Southampton has permitted the continuation of behavioural bioassays begun in the field. In general it has been found that the volatile secretions of predatory ants are repellent to termites, but the non-volatile compounds present on ant cuticle are not.

Work has also begun at Southampton in cooperation with COPR termite group on a comparative study of defensive secretions of soldier castes of Macrotermitinae. Mr P. H. Briner is supplied with study material from the Mokwa project through Mr Longhurst, and from Kenya through Dr Sands' connection with ICIPE. In some genera secretions from both frontal and labral glands are involved, in others, only the labral glands are used.

C LEPIDOPTERA

1 LABORATORY AND FIELD STUDIES OF *SPODOPTERA*

Field tests with the synthetic sex pheromone of the African Armyworm *Spodoptera exempta*

The two components of the synthesised sex pheromone of *Spodoptera exempta* are *cis*-9-tetradecen-1-yl acetate (I) and *cis*-9, *trans*-12-tetradecadien-1-yl acetate (II). Catches of moths of *S. exempta* in pheromone traps baited with the synthetic pheromone were compared with those in light traps. Catches in the pheromone traps consisting of male moths of *S. exempta* were generally lower than those in the light traps although peak catches occurred in both trap systems at the same time in both sites. Factors affecting catches in light and pheromone traps were studied and the use of pheromone traps as an alternative to light traps for use in the East African Armyworm Forecasting system is being considered. The work was in collaboration with P.O. Odiyo, EAAFRO, Nairobi, Kenya and A.M. Mushi, Illonga Research and Training Centre, Ministry of Agriculture, Kilosa, Tanzania.

A project was initiated in Crete for the control of the Egyptian Cotton Leafworm *Spodoptera littoralis* by the use of sex pheromones. The Greek Ministry of Agriculture and the Benaki Phytopathological Institute in Athens welcomed the idea of work on these lines in Crete since in recent years *S. littoralis* has also caused severe crop damage over the whole of Greece, particularly on cotton. Once the problems involved in developing the pheromone as a pest control agent on a small scale have been overcome the method could then be extended over much larger areas of cultivation, particularly on the cotton crop, not only in Greece but also in Egypt and other countries in the Middle East.

The project in Crete was based at the Agricultural Research Station, Ministry of Agriculture, Chania. The programme was to survey the island *Spodoptera* population on the lines previously carried out in Cyprus and also to continue with small scale "communication disruption" experiments using a number of different formulations. Studies on the behaviour of the adult moths in response to pheromone attractants and inhibitors were also continued. Assuming that pheromones can be successfully used against the insect then sustained larval sampling techniques are required for precise estimation of the level of control achieved and studies on these lines were initiated. The behaviour of moths and larvae in the field were also studied.

COPR are also carrying out research on the virus diseases of *S. littoralis* and related species in collaboration with the Unit of Invertebrate Virology (UIV), Oxford with the aim of discovering ways of using the virus to control the insect (see Section 4). One such approach would be to attract male moths by means of a sex attractant to a virus preparation so that the moths would transmit the disease during mating. A search for the presence and distribution of the virus was therefore made. Since the isolation and synthesis of insect sex attractants can involve lengthy and expensive procedures, a number of candidate pheromones were screened to see whether any other important insect pest species were attracted.

Communication disruption between male and female moths of *Spodoptera littoralis* Boisd. was achieved by spraying microencapsulated formulations of pheromone inhibitor (IIA) *cis*-9-tetradecen-1-yl-acetate on lucerne plots located in north-west Crete. Polyurea based microencapsules applied at the rate of 100 g/ha active ingredient caused a 97% disruption for a period of seven days as measured by the reduction of catches in the treated area in traps baited with the pheromone attractant (III) cis-9, trans-11-tetradecadien-1-yl acetate, compared with catches in control areas. At lower rates of application both the level of disruption after seven days and the subsequent persistence of the effects were reduced. When applied at a rate of 100 g/ha active ingredient, disruption of 80% and above was achieved for a period of three weeks. Slower release microencapsules containing IIA based on polyurea/polyamide appeared more effective in that a level of disruption was achieved at one tenth the amount of active ingredient required for the polyurea based formulations.

III dispensed in polythene vials caused 97% disruption when distributed at the rate of one vial per 9 m^2, with lower levels of distruption at distributions of one vial per 25 m^2 or 50 m^2. IIA was much less effective in causing communication disruption at the same rates of distribution and no clear dose/response relationship was established.

Inside enclosures of 10, 100 and 1000 m^2 fenced by a single strand of polythene impregnated with IIA at the rate of 1mg/5 cm and maintained at a height of 0.5 m, a high level of disruption in excess of 93% was achieved for the first five days. The effect persisted in the smaller enclosures in that disruption in excess of 80% was maintained for a period of 40 days. In the 1000 m^2 enclosure the effect declined much more quickly.

When IIA and III were dispensed together in WT traps the inhibitory effect persisted for at least 60 days. Similarly the polyurea based microencapsules containing IIA together with III in WT traps maintained an inhibitory action for periods of 30–40 days. Polyurea/polyamide formulations of IIA were generally less effective. Small numbers of mostly unmated female moths were attracted to the traps when IIA was exposed together with III, either when dispensed in the polythene vials or in microencapsulated formulations.

A number of traps baited with the synthetic pheromone attractant of *Spodoptera littoralis* were distributed throughout Crete. The aim was to find out how extensive the distribution of the insect was in the island and whether semi-isolated populations of the insect existed which could be used for future control experiments using communication disruption techniques. The standard water-traps previously described by Campion *et al.* (1974) were used. The traps were baited with polythene vials containing two mg of the synthetic pheromone attractant (III). A total of 27 traps were used and distributed throughout the island. All the traps were inspected at least once each week, while some were inspected daily, for the period of July to November 1975. The pheromone capsules were replaced approximately every four weeks. Traps located in sites which were visited frequently contained water plus a small quantity of detergent as the catching medium. Traps located in sites visited less frequently contained engine oil and thus avoided the otherwise frequent water replenishment necessary in a hot climate.

No detailed analysis has been attempted at this stage but it is clear that moths are widely distributed throughout the island, although for the most part they occur in the coastal regions. There was no obvious immigration of moths into Crete from elsewhere.

Orientation at close range was seen to differ towards traps baited with pheromone capsules, but staked capsules were also found to be more attractive, at least in the active period after dusk. The capsule on a stake acts as a point source and may easily be located by the male insect; but contained in a trap the capsule produces a very different plume. The airflow around the trap will induce a strongly mixed region inside the trap, containing the pheromone and creating a distributed source which in the close range is confusing to the males. It also seems, however, that there is a long range effect on attractiveness too, which is not clearly understood at the moment but which may be due to the lower mean concentrations to be found in a plume developing from a distributed source (the plume origin is larger and the plume itself will have a considerably larger volume). Trap design experiments carried out underline the difference; a trap designed to represent a point source caught 30% more than the standard. Trap design experiments thus demonstrated that improvements in catch are possible by fairly simple redesign.

More and better detailed measurements of flight tracks in the field are required to define more precisely the flight path, in particular to measure its dimensions under various wind and light conditions. An important factor would be the availability of a suitable night viewing device. The Rank Pullin 'night sight' proved to be good in the field, but it does have serious limitations which, unfortunately, are fundamental to all similar devices.

Pheromone screening for other insects

The isolation and synthesis of species specific insect pheromones may be a costly and laborious process. A considerable number of pheromones of lepidopterous insects have now been identified and with the knowledge of the chemical structure so obtained it is now possible to synthesise closely related compounds in the hope that they may prove to be insect pheromones when tested in the field. The other approach is to expose specific insect pheromones in areas where the insect is not present to find out whether cross-attraction with other species is occuring. The limitation of this approach is that the appropriate insects may not be present in the area or at the time of year that the tests are carried out.

For this series of tests a number of known insect pheromones, single components of multi-component pheromones and other substances of hitherto unknown pheromone activity were tested in the field in Crete.

The 16 candidate pheromones were supplied by Dr Brenda Nesbitt and co-workers of the Tropical Products Institute. The materials were contained in the standard polythene vials at a loading of one mg. They were exposed in WT traps in areas of vegetable and fruit small holdings in north-west Crete. The traps were inspected twice weekly and the catches of moths removed and preserved in 50% aqueous alcohol. After four weeks exposure the capsules were renewed for a further four weeks. The tests were carried out during the months of September and October.

None of the candidate pheromones attracted an appreciable number of any particular moth species during the test period of eight weeks. Cylohexane ethane carboxylate, a substance with a similar I.R. absorption to the pheromone attractant of *S. littoralis* and which therefore according to Wright should also be attractive, caught nothing.

Virus studies

Larvae were sampled from throughout different parts of Crete. Virus outbreaks were noted in the lucerne growing areas of north-west Crete. Subsequent tests at UIV, Oxford confirmed that this was the same strain as that at present being safety tested.

Sampling techniques

As part of the virus investigation an extensive search was made for the eggs of *S. littoralis*. Very few were laid on crop plants. The majority of egg masses were found on the giant grass *Arundo donax* L. that grows wild in Crete and is used as windbreaks in cultivated areas. Further extensive searches for eggs in the course of this activity confirmed this conclusion. The egg batches are easy to see on the broad leaves and stems of the grass and it is possible to write on the leaves with a felt tip pen, so that batches can be marked and indexed. Remains of egg batches are visible on the reeds for many days after hatching.

On two occasions windbreaks were sampled by examining three non-flowering and three flowering stems of the giant grass (flowering stems are generally taller and have larger leaves) every ten paces. On the first occasion 120 m of break were sampled in one hour; 39 vegetative and 39 flowering grass stems were examined. Eight batches of unhatched eggs and seven batches of hatched eggs were found. On the second occasion 60 m of break were examined in 20 min, but no eggs were found. It would seem reasonable to assume that 100 m of windbreak could be sampled in one hour by one person. Although it is necessary to pull the grass stems down to eye level for examination, this can usually be done without damage to the plant. Egg sampling would appear to be simple and rapid to carry out, to do little damage to farm crops and to require only a small amount of labour. In one large cultivated area at Drapanias 456 m of windbreak were present in 138 000 m^2 of cultivation. It would not seem unreasonable to expect one person to be able to sample these windbreaks in five hours. The total cultivated area at Drapanias is some 70 ha so it may be possible to sample the whole of this area for eggs in some 25 man-hours.

Larvae were sampled on five different sites. It was necessary to make excursions into the lucerne so that some trampling of the crop was unavoidable. Especially during the hotter part of the day a proportion of the larvae hide in the trash at the bottom of the plants, whilst others may hide in cracks in the soil. The larvae were therefore collected and counted as follows. After the area had been pegged out with sticks, the larvae were picked off the taller plants. This was followed by a more thorough search of the stems down to ground level, followed by a search of the ground, trash and in the soil cracks.

Patches of lucerne are small (500–1000 m^2) and samples were taken at regular intervals through the crop. It is of course only possible to sample a very small proportion of each field and it is hoped that statistical analysis will help to give guidelines about the porportion of an area that should be examined in order to obtain a reasonable population estimate.

In spite of several night-time searches for moths, few were seen. Only two moths were seen amongst the lucerne and both of these were newly emerged females, although during studies on virus, numerous pupae were unearthed in lucerne fields. Males were also seen flying low over the lucerne towards traps baited with sex pheromone. A few females were found resting and laying eggs on the leaves of the giant grass, *Arundo donax*, whilst feeding and copulating males and females were found on a tamarisk tree that was in flower. "Calling" females (the pheromone gland is exposed) were only found in the tamarisk tree. The night-time resting places of the moths may be governed by three factors. Firstly, there was a tendency to settle on vegetation that was above 1.5 m from the ground. Secondly, there was a tendency to remain on the vegetation that had provided food. At the time of the lucerne flowering more moths may be found in the crop. Thirdly, certain vegetation

Fig. 16. A miniature meteorological station set up in a field in Crete during the armyworm control experiments.

Fig. 17.
A Kytoon being raised with insect release gear. The released insects are to be followed in a radar tracking experiment.

may be attractive to females for egg-laying. Should it prove to be a general rule that "calling" females are found on the taller plants surrounding the crops, then this may have to be taken into account in trials involving communication disruption.

Feeding studies

Although a polyphagous insect, *S. littoralis* larvae cannot survive on all plants. In Crete, lucerne is obviously their main food, although vegetables like tomatoes, potatoes, beans and cabbage sometimes provide alternatives. In tests carried out in Crete, a number of plants found growing wild in the fields were capable of supporting the larvae for a few days. These included various species of grass and brambles as well as the wild *Rumex* and *Chenopodium*, plants that are capable of sustaining the larvae throughout life. *S. littoralis* is not generally considered to be a grass feeder and larval life cannot be completed on such a diet. With the Crete strain some 15% of those fed on maize completed larval life and pupated, but the resulting adults were very small in size. During the virus studies some larvae and pupae were found in maize fields.

Tests in Crete also included ones in which small groups of larvae were offered a choice of fresh lucerne and one other wild plant. In all cases the majority of larvae moved from the test plant to the lucerne. Thus over 90% did so when the alternative was a grass or bramble or a common weed species of *Senecio*; 60—70% when the alternative was a species of *Chenopodium*.

Only when the alternative was a species of *Rumex* did more than 50% of the larvae remain on the alternative. Blanket treatment of whole areas by communication disruption chemicals may therefore be necessary for maximum control efficiency.

Behaviour in response to pheromone

Problems studied relating to the behaviour of *S. littoralis* moths in response to a plume of sex pheromone included:
1. The period of activity of males responding to pheromone stimulus and the distribution of arrivals at the source during this active period.
2. The effect of wind speed and direction on the arrival distribution.
3. The flight path of males responding to pheromone stimulation (observations and measurements).
4. The evaluation of an image intensifier (night sight) intended for entomological field work.
5. The effects of natural inhibiting agents on male flight behaviour.
6. Experiments with the design of pheromone traps.

It was shown that the broad casting-type movement and the shallow zig-zag are short range responses and that at long range the flight is straight and relatively fast. It also seems that casting or hunting occurs in the vertical plane with horizontal movement being very limited compared with vertical oscillations.

The pattern of arrivals, clusters just after a gust, suggests that flight towards a pheromone source may be made in several stages either during or just after a gust, the male alighting once the stimulus has worn off.

2 BIOGEOGRAPHICAL STUDIES OF THE AFRICAN ARMYWORM, *SPODOPTERA EXEMPTA*

Recorded infestations of the African armyworm in East Africa during 1975 occurred over areas totalling some 900 km^2, an order of magnitude less than was infested in 1974. Unusual features of the season were the abnormally late development in June of serious infestations in the Kenya highlands and the appearance of two subsequent generations of larvae farther north in Kenya in Laikipia and Samburu in July and in Turkana in August. Emigration from East Africa to Ethiopia in May—July appears to have been on a restricted scale only and no infestations were reported in Yemen Arab Republic. This is in marked contrast to the previous season when massive infestations extended progressively from Tanzania through Uganda and Kenya to Ethiopia and south-western Arabia.

A preliminary paper on these extensive and large-scale migrations of *Spodoptera exempta* moths in 1974 by Miss E. Betts was sent to the armyworm Workshop at the International Centre for Insect Physiology and Ecology in Nairobi, and further analysis of the material has continued.

Notes, maps and graphs of the seasonal and annual changes in distribution of African armyworms in Eastern Africa over the last 14 years and the implications of the changes for formulating an effective strategy of armyworm control were sent to DLCOEA. Assessments of major developments in the current armyworm situation continued and on four occasions armyworm reports and warnings of possible developments were passed on to entomologists in Ethiopia, Yemen Arab Republic and Sudan and East Africa. Liaison continued with the Armyworm Forecasting Service for East Africa, with the Entomologist at the Institute of Agricultural Research in Ethiopia and with armyworm scientists of the International Centre for Insect Physiology and Ecology.

The paper on the Genus *Spodoptera* in Africa and the Near East by the late E. S. Brown and C. F. Dewhurst was published in *Bull. ent. Res.*

3 *CHILO PARTELLUS* OVIPOSITION BEHAVIOUR

Early in the year work was continued on addition of maize seedling leaf powder to the artificial diet. Following a previous report and comments by Dr Chapman this work has been taken up again and is being carried out by a visiting student. He is comparing (1) the effects of rearing for various periods on maize seedling leaves before transfer to artificial diet with continuous rearing on artificial diet and (2) the effects of adding lyophilised leaves of various ages, lyophilised stems and lyophilised wheat leaves to the diet with rearing on ordinary diet. Larval development, survival and subsequent adult morphology, fecundity and mating ability will be recorded and compared. The student is also carrying out a study on post-eclosion behaviour of young larvae.

Dr Roome is continuing with the oviposition preference tests. Further plant leaves have been compared with maize seedling leaves as oviposition sites. None has been preferred to maize. Hairiness is still a common factor among distinctly non-preferred leaves. Only a smooth leaved plant called "rocket" (*Eruca sativa*) showed reduced oviposition which could not be most easily ascribed to physical factors. Older (10+ weeks) maize leaves have a distinctly hairy upper surface. Maize leaves become less preferred as hairiness develops, starting around leaf 8. The effect of aging in maize on oviposition is still being examined. Agar gel may provide a medium in which the ability to discriminate chemical differences may be tested. Oviposition occurs on it in a stainless steel gauze cage and *Chilo* can distinguish between saline (7%) agar and non-saline agar. This method will be used as far as possible to show discrimination (1) of pure chemicals (2) of plant extracts, and to reveal the role played by the various sense organs in this discrimination. The direct observations made by low-light surveillance instrument on oviposition behaviour suggest that the antennae and ovipositor play active roles, and it is possible that the tarsi are also involved. Mrs Chadha's work on the ovipositor of *Chilo* has confirmed the presence of two pairs of chemo-receptor hairs. Behavioural evidence of the role of these hairs is now required, and if this is obtained, Dr W. Blaney of Birkbeck College has agreed to examine the hairs using electro-physiological methods. The behavioural evidence may be obtained by combination of the agar gel test with extirpation of the sensillae and Dr Roome will be undertaking this work. Mrs Chadha has prepared ovipositor tips of *Diatraea* sp., and *Margarodes* sp. (from Dr King at Turrialba) for scanning electromicroscopy. These are both borers, one of monocots (sugar-cane) and one of dicots (sweet-potato) and it is thought that a comparison with the ovipositor of *Chilo* will prove rewarding. Other than attempts to obtain a good scanning photograph of the sensillum pore, and a little further T.E.M. work, the ovipositor work is now over and Mrs Chadha is examining the antennae and tarsi with S.E.M. before going on to look at larval sensillae.

4 INSECT VIRUSES

Work on the joint COPR/NERC virus research project continued with particular emphasis on the virus of *Spodoptera littoralis* as being the most likely first candidate for field testing.

The bioassay methods developed previously were used extensively during the present year. The precise bioassay was used to build up an increasingly detailed picture of the variation of LD50 with age, extending the procedure both to very young larvae and to those nearing pupation. Bioassays on successive days on samples of larvae from a single batch gave a clearer picture of the periods of increased susceptibility that occur at about the time of moult. Crude bioassay was used to compare different batches of virus, the effect of hypochlorite and other disinfectants commonly used in the laboratory, and a number of potential spray additives including buffer, molasses, polyethylene glycol and sucrose. Only sucrose had the effect of decreasing the infectivity of the virus and this primarily because of its effect on the larval appetite.

A long term experiment comparing the activity of virus stored in water at room temperature and that kept in the deep freeze showed no difference after eight months. Work has started using storage temperatures likely to be met with in the tropics. A technique is being developed for screening dyes and other materials with potential as ultra-violet light protectants, and will be used overseas to determine the effect of natural sunlight.

A unique feature of the project has always been that virus samples obtained from sufficiently large bioassays or from the field can be positively identified by the staff of the Unit of Invertebrate Virology, NERC. Having previously determined the LD50 for *Spodoptera littoralis* and *S. frugiperda* at different ages with their respective homologous viruses, some bioassays were repeated on a larger scale, and where susceptibility was close to that previously determined the dead insects were sent for identification of the virus. In all cases the virus was found to be identical with that fed to the insects. Studies carried out last year treating insects with non-homologous virus fed at very high doses seemed to show cross-infectivity, and LD50s were determined.

Repetition of these tests with identification however has so far shown death from the insect's own (homologous) virus except in one instance where seven-day-old *S. frugiperda* larvae fed with a high dose of *littoralis* NPV died and yielded *littoralis* virus. Investigation of these peculiar results is being carried out.

Spodoptera littoralis virus was generated as an aerosol with droplets of between one and ten microns diameter at 40% RH as part of the MRE safety testing programme and was used for bioassay using first instar *S. littoralis* larvae. Conversion into an aerosol and collection in water had no effect on viral infectivity compared with material taken from the generating vessel. It was similarly found that the virus was not affected by freeze drying, nor by drying on glass at room temperature and resuspending in water.

With the installation of a Porton Safety Cabinet, work with pure freeze-dried virus dust became possible. Application of 10% virus in talc to male pupae had no effect on first instar larvae produced by mating the emerging moths with virgin females. Male moths dusted with the same material and allowed to feed and fly for 24 hours with two changes of cage before mating with females from outside the test area produced larvae showing over 90% deaths from virus in the first five days after hatch (as compared with no deaths resulting from matings with control moths kept in the safety cabinet, though in separate cages, with the infected moths).

A field trip was made to S. Tanzania during January, following the Armyworm workshop at ICIPE in Nairobi. *S. exempta* collected on this trip at sites further south than any previous NPV outbreak in E. Africa yielded virus but this awaits bioassay and characterization due to temporary shortage of live insects. A more extended trip was made to Crete in September to observe the lucerne crop which was heavily infested with *Spodoptera littoralis*. Two viral outbreaks were observed in detail, and samples collected were identified by UIV as being identical with the "Oxford" virus used in the laboratory work. The same material was also identified in larvae collected in an apparently healthy state from lucerne, in larvae wandering between fields, and in larvae reared from eggs collected from giant grasses (*Donax* sp.), all of which were stressed after collection both by deliberate overcrowding and partial starvation, and by 48 hours cold. All larvae which took longer than five days to die after being stressed were found to have died from causes other than virus. Bird droppings collected in the same area are being examined. Those from the area of the virus outbreak were examined microscopically and many were found to contain a NPV with the same physical characteristics as the "Oxford" virus. Samples of these viruses were then fed to larvae from the Porton colony and the larvae died of virus which was shown to be identical to the "Oxford" strain. Without collecting enormous numbers of bird droppings to enable direct identification of the polyhedra present this work gives the best available confirmation that viable *S. littoralis* NPV was being spread by birds in Crete. Soil samples taken from a number of sites in which the upper few millimetres were first removed have failed to yield virus polyhedra even when taken from lucerne fields containing many virus-infected larvae.

D METEOROLOGICAL AND RADAR STUDIES

1 METEOROLOGY

Mr D. E. Pedgley spent part of his time during the year as a Rapporteur to the World Meteorological Organization, preparing a draft report on the meteorological aspects of aerobiology, suitable for publication as one of the Organization's Technical Notes. The aim is to give a concise account of how atmospheric behaviour affects the behaviour of airborne organisms. This account should serve two main functions: to help aerobiologists understand how the atmosphere works (in particular, the types and origins of wind systems on all scales), and to help meteorologists understand how organisms behave (in particular, how they are taken into, with, through and from the air). It could help the planning of more interdisciplinary approaches to problems of controlling the spread of airborne pests and diseases, emphasising particularly the need to examine any role there may be for airborne spread of a given pest or disease, including the need to set up adequate sampling networks for both sources and cloud structures of the organism, so that control strategies can be devised in the light of the organism's behaviour. The report gives, in outline only, what is known about atmospheric behaviour in relation to organisms in it, together with numerous examples from the literature of the wide variety of organisms, scales of atmospheric disturbance, geographical locations and experimental methods that have been studied. It should be finished during 1976.

For six weeks, 6 April–16 May, Mr Pedgley was in Ougadougou, Upper Volta, as part of the COPR contribution to the WHO/OCP. His work was to examine the weather patterns on certain days when flies first started to bite at wet-season breeding sites, with the aim of seeing if there were any similarities that might suggest fly movement downwind. Movements over distances of 100 km have already been suggested; if they take place frequently they would need to be considered in developing the strategy of fly control. Preliminary findings, based on the few cases available for study, show that flies first bite, i.e. they reinvade (assuming they do not simply stay put without biting during the dry season, and they bite soon after arriving), close to well-known types of wind convergence — squalls and inter-tropical convergence zone. In all cases flies reappeared only after re-establishment of the monsoon wind flow, suggesting a source somewhere to the east, south or west. Similar movements are suggested by reinvasions in 1975 of areas controlled by the OCP. A better understanding of the ranges and frequencies of possible airborne movements will need more adequate sampling networks (for both flies and their breeding sites), and a better understanding of flight behaviour.

Whilst in Upper Volta, Mr Pedgley also gathered data for a detailed study, now started, on one particular line squall, to examine its winds as potential carriers of flies.

A survey of summer winds in the Southern Red Sea was completed by Mr M. R. Tucker. A well-defined convergence zone was found between south-west monsoon winds over the Gulf of Aden and southern Ethiopia and northerly winds over the southern Red Sea and northern Ethiopia. The northerly winds come partly from the northern Red Sea and partly from Sudan, south of the Inter-tropical convergence zone (ITCZ). The 'Afar convergence zone' resembled both the ITCZ and a lee convergence, where two branches of the south-west monsoon meet after flowing around opposite sides of the Ethiopian Highlands.

Fig. 18. One of the hazards of operating in remote areas — the radar vehicle about to cross a river on a pontoon.

Fig. 19. The radar set up for action during field work in Mali

An analysis of rainfall records is in progress for five meteorological stations close to the River Niger in the sahel zone of West Africa. Monthly rainfall totals and numbers of raindays are being analysed from the time of earliest records (1908 for Niamey) up to 1974. Five-year running means are being used to analyse rainfall trends and to find periodic variations in rainfall totals. They are compared for different stations and will be used in conjunction with the results of grasshopper surveys in the area. A survey of the literature on the drought in the West African Sahel has also been started.

2 RADAR

Detailed analysis of data accumulated in Mali from October–November 1974 has yielded quantitative descriptions of insect flight behaviour in the area of the Niger flood plain within a few kilometres of Kara (14°9'N, 5°3'W). Displacement was generally to the south-west, i.e. with the prevailing Harmattan wind. Extrapolation of flight trajectories beyond the range of radar indicated that in certain weather conditions overnight flight by insects (possibly small grasshoppers) may exceed 350 km. On nights of light winds the insects often headed into the wind producing a northerly displacement.

In order to provide a quantitative description of the collective orientation phenomenon, detailed analyses have been undertaken of the bias effects introduced in trajectories displayed by the PPI scanning system. These effects were found to be substantial and a knowledge of the angular variation of radar cross section of the targets of interest proved necessary for rigorous elimination of the bias.

Further signature analysis confirmed earlier predictions that identification techniques based on wingbeat frequency alone did not provide enough information for reliable species recognition, at least in the complex aerial fauna found in the flood plain. The signature information did however set an upper boundary to the percentage of *Locusta* amongst the larger (mass > 0.1 g) aerial fauna. In the altitude range 30 m–1000 m this upper boundary was $\sim 15\%$, the actual percentage of *Locusta* probably being very much lower.

Improvements in the identification capabilities of the radar have been sought by the development of a prototype vertical-looking system in which the plane of polarisation of a circular symmetric beam is continuously rotated. The variation of magnitude of the echoes received from targets passing through the beam was expected to provide information about the target body shape, as well as details of the wingbeat pattern. Separation of targets with overlapping wingbeat frequencies but different body shapes should thus be possible with this system. Initial examination of data in Mali in the autumn of 1975 indicated that the technique produced results of the type expected.

An alternative approach to the identification problem is to select areas of simple aerial fauna for radar operation. This was also attempted in Mali, by taking a "simple" radar system to the northern lake bed area (15°47'N, 3°2'W). The preliminary results from this equipment were also encouraging.

3 SUDAN GEZIRA PROJECT

For various reasons, COPR staff were not involved in the aerial field work in the Sudan during 1975 as in previous years, but at the request of the Sudan authorities,

Dr Rainey spent 5 weeks in the Sudan working on the establishment of a system at the Sudan Gezira Board headquarters for the mapping and analysis of current Gezira pest survey data and of relevant environmental factors to assist in the direction and supervision of pest control operations. SGB staff were introduced to the methods of mapping and analysis developed by Miss Haggis in the course of earlier seasons' work and with the joint research project directed by Mr R. J. V. Joyce of the Ciba-Geigy Agricultural Aviation Research Unit, and initial arrangements were made for continuing this training programme at COPR.

E CROP PROTECTION PROJECTS

1 CROP PROTECTION OPERATIONS IN THE SAHEL

After the period of disastrous drought in the Sahel, attention was turned to more long-term projects to assist the agricultural recovery of the area. In order to facilitate these activities, the FAO Office of Sahelian Relief Operations established a coordinating service, supervised by three staff members detached from COPR: Mr G. B. Popov, Dr M. R. K. Lambert and Miss P. McAleer, who in close collaboration with the national plant protection services of the Sahelian states and the two regional locust control organisations, OCLALAV and OICMA, helped to organise and conduct the control against such crop pests as grasshoppers, caterpillars, including various species of stemborers, and some plant bugs, aphids and beetles. Insecticides, control equipment and vehicles worth over five million U.S. dollars and an additional quantity of transport and equipment were distributed by OSRO throughout the Sahelian Zone and despite the serious problems of communication and transport during the rains, most of it found its way in good time to the front line of the Control Units and the hands of the farmers themselves. Although the pest attacks were heavier even than the most pessimistic forecasts, it was possible to avert the worst of the damage by controlling a total area of over one million hectares, about half of it controlled by ground teams and farmers and the other half by aerial spraying. This is the first time in history that the control of subsistence crops has been undertaken on such a scale in the Sahel and it has done much to relieve the famine conditions there. In turn the success of these operations owes much to the many activities performed as part of the coordinating service in the following fields: the initial planning of operations and the definition of spheres of action between the various control units; special surveys to locate assess and delimit the infestations; timely warning of serious developments requiring emergency intervention; the planning of day to day strategy of central operations and development forces; training of technical personnel of plant protection services and of farmers in the handling, preparation, application and storage of bait and insecticides and the use of various types of control equipment; monitoring of the crop pest situation and the activity of various pests and reporting on the situation to OSRO and the various bodies concerned.

Monitoring surveys

In order to keep track of the pest situation in time and space, a series of monitoring stations was established along a transect crossing the Sahelian zone from its northern extremity at Gao with a rainfall of about 250 mm and following the course of the river Niger across the Sahel southwards to the Niger/Dahomey border at Gaya—Malanville, where rainfall is about 750 mm. The siting of stations along the Gao—Gaya road ensured easy communication and ready access to the river valley with its wide range of ecological conditions, types of crops and crop pests.

Each station was manned by a technical officer assigned from a national or regional plant protection service who was responsible for periodical field surveys of crop pests at chosen sites in proximity to his station and for conducting light trap and weather observations. This work was supervised by Dr Lambert of COPR and Mr T. Yonli, a scientist from OCLALAV and also by Mr Popov, during his occasional visits.

The results are of considerable interest in throwing much light on the economic importance of the various insect pests. Grasshoppers are the most important group of crop pests, responsible for considerable damage to practically every type of crop, especially to cereals such as millets, maize and sorghum. Although the number of species is greater in the south, the damage is even higher in the northern part of the Sahel, because the fewer species there tend to multiply to higher numbers, often reaching swarming proportions. Many species are highly mobile and fly by day, or more often by night singly or in concentrations, over distances of tens or even hundreds of kilometres. Evidence from light trap and field observations suggests that such movements have a seasonal pattern associated with the dynamics of the Inter-Tropical Convergence Zone (ITCZ) which also bring the monsoon rain to the Sahel. As an example, the Senegalese grasshopper (*Oedaleus senegalensis*), economically probably the most important species, appeared in large numbers from the eggs remaining in the ground from the previous season at the beginning of the rains at a number of points in the southern part of the Sahelian zone. The new adults appearing in July moved northwards to the northern part of the Sahel where they bred in the wake of the rain. Two or possibly three generations were produced and the resultant adults moved back with the retreating ITCZ laying on their way to the southern part of the Sahel.

In addition to grasshoppers various other insects, particularly stemborers, caterpillars of noctuid moths, sucking bugs like *Dysdercus* species and a number of vertebrate pests, especially weaver birds, *Quelea* species and the Golden Sparrow among them, and various rodents, were responsible for much damage, perhaps as much as 25—30% over much of northern Sahel and locally resulting in complete destruction of the crop. In the south the damage was on the whole lighter.

Other studies

Miss McAleer assisted with the analysis of the results of the work of the monitoring surveys, particularly in relation to the weather factors — rainfall, wind, temperature — likely to control the flight of insect pests. She also conducted rearing experiments with some species of grasshoppers as a study of their biology and life cycle.

In addition to their immediate interest for the OSRO Emergency Campaign, these studies are a worthwhile contribution to a better strategy by improved knowledge of the biology, ecology and seasonal breeding and movements of some crop pests.

2 REGIONAL PLANT PROTECTION TEAM IN THE CARIBBEAN

One of the major objectives of the COPR team stationed since late 1974 in the West Indies under UKTC is to develop suitable methods for the control of insects on Sea Island cotton, recently much increased in importance. The work in Barbados is carried out by Mr W. R. Ingram (team leader) and in Antigua by Mr S. N. Irving.

There is no shortage of pest problems the most important of which seem to be the armyworms *Alabama argilacae* and no less than five species of *Spodoptera*. In Antigua they are joined by *Heliothis virescens* and *H. zea* and also by *Dysdercus anchea*, while in Barbados *Bucculatrix thurbiella,* another defoliating moth, is of

Fig. 20.
Pink bollworm larvae on a cotton boll in the West Indies.

Fig. 21. The Irving-Jackson ULV TM Sprayer spraying a cotton crop in Antigua.

sporadic importance. *Pectinophora gossypiella* is held in check by a close season of three months. In addition there are many other pests of relatively minor importance.

The team has put its main effort into ULV application. They were highly successful with *Alabama*, which occurs on the upper side of the leaf on top of the plant, but *Spodoptera* has so far proved a very difficult problem as the spray does not seem to reach the underside of the leaves lower down on the plant.

Dr G. A. Mitchell, the third member of the COPR team, is working at the Winban Research Station in St. Lucia. Apart from bananas, he takes an interest in other cash crops grown in the four islands supporting Winban, namely St. Lucia, Dominica, Grenada and St. Vincent.

On the entomological side, the biggest problem in bananas remains the weevil *Cosmopolites sordidus*. This used to be controlled satisfactorily with aldrin or heptachlor, but this is no longer effective. Resistance has arisen in some places, particularly St. Vincent, but the majority of failures arise from bad application or no application at all. As the aldrin/heptachlor treatment is only about a tenth of the cost of any practicable alternative, it should be continued where the insect is still susceptible. Elsewhere chlordecone dust (Kepone) or pyrimiphos-ethyl spray (Primicid) could be used, but the latter requires a new technology and weed control, therefore close supervision.

Other damage is cosmetic by thrips and crickets. It does not affect the quality of the fruit but severely restricts its marketability. At present this is controlled by sleeving the bunches in polythene bags impregnated with chlorpyrifos (Dursban), but sprays, including permethrene, will be tried.

Mr W. R. Ingram also paid two visits to Grenada during the year to advise on coconuts and cocoa. Coconuts were found to be heavily infested with the mite *Eriophyes guerreronis* and arrangements were made for a visit by Dr N. W. Hussey of GCRI to advise on possible biological control. The spraying system on cocoa for control of *Steirastoma brave*, thrips, scales, mealybugs, whiteflies and blackpod (*Phytophthora palmivora*) was considered to be wasteful. An advisory visit by Dr G. A. Matthews of OSMC to advise on improvements was arranged and an experimental programme to identify possible alternatives to dieldrin was planned.

3 MULTICROPPING SYSTEMS RESEARCH IN COSTA RICA

This very promising research project in progress at the Tropical Agricultural Research and Training Centre (CATIE), Turrialba is designed to bring existing technology within the means of the peasant cultivator, who supplies a large part of the food consumed in Central America. The project is now in its third season, and on a central experimental site on the land of CATIE no less than 54 different systems, involving combinations of maize, sweet potato, beans, upland rice and cassava are treated as monoculture, mixed cropping and intercropping with different degrees of overlapping and sequence. On a somewhat smaller scale, the work is repeated at San Isidro on the Pacific side of Costa Rica where climate, soils and crops are different.

Dr A. B. S. King, a COPR entomologist, was seconded to the project under UK Technical Cooperation from June 1975. At present his major task is to assess the relative importance of the various insect pests. To this end he is monitoring by light traps and other means the insects occuring in the central plot which is treated regularly with insecticides, comparing them with what he finds in two isolated situations nearby, planted for the purpose with maize and sweet potatoes, one in alternate rows, the other in separate plots. He is also trying to establish damage levels and yield loss by artificial infestation of *Megastes rundella*, one of the most important pests on sweet potato. He measures activity and oviposition cycles of *Megastes* and *Agrotis* which he finds depend on light intensity, being reduced at full moon. This is done in the field as well as in cage experiments, and will be extended in due course. He also hopes to start soon on control experiments by simple, safe and cheap methods including ULV.

F PUBLIC HEALTH PROJECTS

1 SCHISTOSOMIASIS

Introduction

In common with many other areas of pesticide work, more data is now being required of those involved in the chemical control of the snail intermediate hosts of schistosomiasis on the possible acute and long-term effects of the use of molluscicides on the environment. The COPR Molluscicides Group is contributing information on this aspect by concentrating its efforts firstly on the effects of molluscicides on non-target organisms. Up till now, this has involved work on fish but it is hoped that the laboratory studies to date will be added to this coming year with the start of a field project in Ghana. This aims to look at acute and chronic effects of chemicals on a much wider range of fauna and flora in self-sustaining fish ponds. Secondly, the mode of action of molluscicides is being studied by a variety of radiochemical methods in order to see how uptake into and distribution within snail and fish tissues may compare with other 'non-molluscicidal' compounds. Attention is being particularly focussed on a group of chemically related nicotinanilides since they appear, at least, to have the merit of no acute toxicity to fish at molluscicidal concentrations.

Field work in Ethiopia has now been terminated after a period of eight years during which COPR staff contributed to studies on the epidemiology and control of schistosomiasis in the Awash Valley and the township of Adwa in the highlands. Present field efforts are concentrated in The Gambia with a study of the longevity of the *Schistosoma haematobium* adult female parasite, the control of transmission from seasonal pools and the possible effect of increasing rice culture on transmission. In addition to the above mentioned Ghana project there is the possibility of work on schistosomiasis control starting in Swaziland next year; these various projects representing an increasing emphasis on field work within the Group.

The synthesis of radioactively labelled molluscicides

The determination of the rate of uptake of molluscicide by both snails and fish has required the synthesis of radiochemically labelled compounds. Nicotinanilides, labelled with tritium in the aniline ring have been made. In addition, dual labelled nicotinanilide (carbon-14 and tritium) has also been synthesised and has been used to study its metabolic breakdown in *Biomphalaria glabrata*.

A comprehensive series of substituted nicotinanilides has been prepared and the chemical and physical properties of these compounds studied using mainly spectroscopic methods. The results are being compared with measured LC50 values against *B. glabrata*. The estimation of partition coefficients is a necessary step in attempting to explain the biological activity of a related group of pesticides and a relatively simple method, based on reverse phase chromatography, has been developed for the purpose.

Hydrolysis of the nicotinanilides takes place slowly in water and the rate of hydrolysis of 4'-chloronicotinanilide has been determined at biologically active concentrations using a tritium labelled compound. It was found to have a half-life of 8 to 10 days. When stored as solids or in ethanolic solution the nicotinanilides are quite stable. They also appear to be stable to ultraviolet irradiation. Knowledge of these factors is a prerequisite to future field use of these compounds.

Studies on the uptake of molluscicides by snails

Earlier work on the uptake of molluscicides, based on exposure of snails to relatively small volumes of solution in beakers, has shown that a straight line relationship exists between log shell size and log amount taken up in a set time. The method, however, has obvious limitations. Even with quite short exposures, the animals tend to crawl out of the medium. This may be attributed to either falling oxygen tension or increasing pollution by excretory products. An alternative approach has, therefore, been sought and a flowing water system developed. The apparatus for this consists of a cylindrical glass cell, 5 cm high by 2·5 cm diameter, into which water may be passed at a constant rate by air displacement from a Winchester bottle by means of a peristaltic pump. A flow rate of 1 ml/min is usually employed. Groups of seven snails are allowed an acclimatisation period of two hours after which a three-way tap is used to redirect the flow to a second Winchester containing isotopically labelled pesticide solution. A higher rate of pumping for a few minutes at this time suffices to completely exchange the water in the cell. The cell contents are evacuated by an arrangement which provides for an even cross-sectional flow throughout the length of the cell, removes faeces and any other secreted or excreted products from the ambience of the snails and maintains the water level constant in the cell. The cell system is replicated ten times and the entire arrangement is held in a large water bath at 24°C. The system will run continuously for up to five days and the snails remain immersed in the cell during this time even without feeding.

Uptake appears to be affected by activity. Snails do have higher uptake rates when initially introduced into a medium than after a few hours of acclimatisation. That activity declines to a steady value with time of acclimatisation has been confirmed by time-lapse photography. In addition, heart rate, which has been shown to have a negative asymptotic relationship with shell size, can also be shown to fall during the acclimatisation period. Heart rate also appears to be affected by activity therefore. That uptake and activity are related may be explicable in terms of the amount of surface area of the head-foot exposed during crawling as compared with a more sessile condition and that uptake depends on the lipid constituents of the external membranes of the snail.

Comparisons have been made between the uptake rates of N-tritylmorpholine, morpholine and three para-substituted nicotinanilides. It is hoped to show that the uptake is related to the lipophilic character of the compounds and, therefore, to the Hansch π value. In order to further elucidate the site(s) of entry of any compound, studies have been made on the influx rate of water to see whether this has any influence on the uptake of other compounds.

Autoradiography

Autoradiography is being used to study the sites of uptake and distribution of labelled molluscicides in the snail hosts of schistosomiasis. Autoradiographs have been prepared from *B. glabrata* previously exposed to either N-tritylmorpholine-H3 or 4'-chloronicotinanilide-H3 at sublethal concentrations for 24 hours. It first appeared that the molluscicide, as indicated by the presence of silver grains in the photographic emulsion, is distributed extensively throughout the snail. It was later found, however, that the true distribution is obscured by artefacts due to the chemical action of oxidising and reducing agents in the tissue.

In autoradiographs prepared from frozen sections of *B. glabrata*, negative chemography is associated with the haemolymph and positive chemography with the

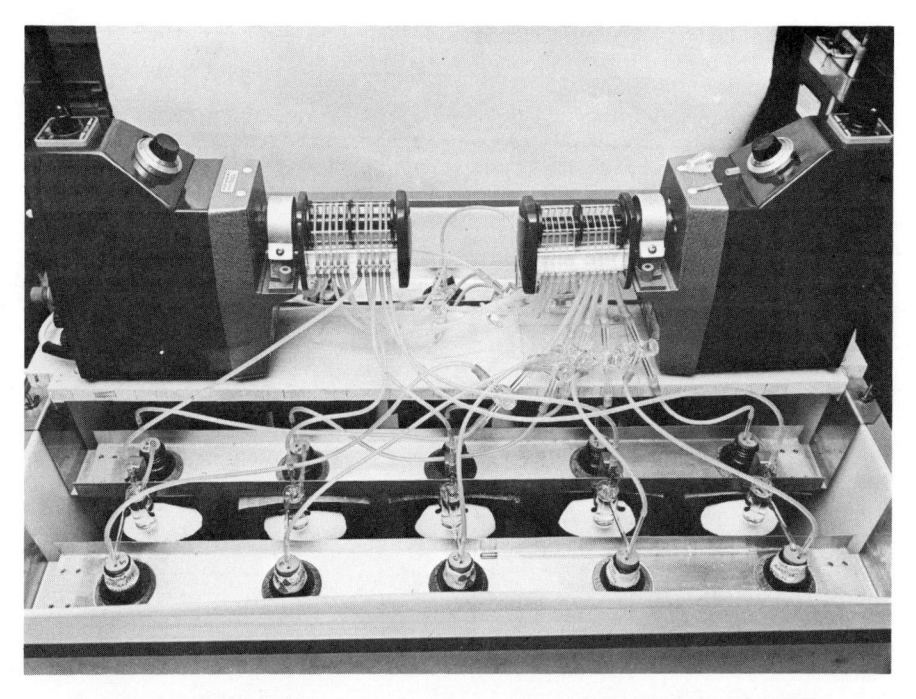

Fig. 22. Apparatus for the exposure of aquatic snails to pesticides under flowing water conditions.

Fig. 23. Small specimens of *Sarotherodon mossambicus,* an important food fish, bred in the laboratory for testing the effects of pesticides.

prostate, the radular sac, the mucus secreted by the edge of the mantle and, to a lesser extent, the cells of the oviduct and digestive gland. A high degree of negative chemography was to be found associated with the yolk in frozen sections of newly laid eggs. In sections of 3 day-old eggs, the negative chemography is less evident and in sections of eggs just prior to hatching, it is restricted to around the digestive glands of the fully developed ova. Positive chemography was only observed in the fully developed embryos although in earlier stages of development it may have been erased by negative chemography.

To overcome these effects, the snail sections and emulsion are now placed on opposite sides of a 170–210 mm thick polyvinylchloride membrane during the exposure period. This eliminates chemography and the membrane is thin enough to allow the passage of β particles. The quality of the sections, however, still varies considerably and alternative techniques involving freeze-dried sections are, therefore, being investigated.

Fish toxicity studies

Laboratory work on the toxicity of aquatic pesticides to the tropical food fish, *Sarotherodon mossambicus* has been continued by Dr P. Matthiessen and C. M. Fox with the aid of a WHO grant. Emphasis has been placed on modes of uptake and excretion of these substances. The manufacturer of the slow-release formulation tributyltin oxide, which previous studies appear to have shown to have deleterious side effects on fish (see COPR Report 1974) has recently decided to curtail its further development. The much more promising molluscicide, 4'-chloronicotinanilide (DOWCO 215), which we have shown to be relatively harmless to fish at molluscicidal dosages unfortunately could not be investigated further in 1975 since the chemical is no longer being produced by the patent-holders (Dow Chemicals Inc.). It is hoped that pilot plant quantities of DOWCO 215 can be made and field-tested in connection with the forthcoming COPR aquatic pesticides project in Ghana described below.

Uptake, metabolism and excretion by *S. mossambicus* of the well-established molluscicide, N-tritylmorpholine (Frescon, Shell Chemicals) has been investigated with the aid of the radiolabelled compound. Preliminary tests showed that the strain of *S. mossambicus* used is quite susceptible to Frescon (24 hr LC50 = 0.017 mg/litre) so all uptake experiments were performed at lower, sublethal concentrations (0.005–0.010 mg/litre). It was found that both adults and fry accumulate Frescon at a high rate, that approximately 80% of the absorbed compound remains chemically unchanged and that loss rates are relatively slow. After 10 hours exposure, the concentration factor (c.f.) of the fish tissue with respect to the environment is already x 60 and by the 55 hour point, the c.f. reaches x 1200. This means that fish exposed to a molluscicidal dosage (0.025 mg/litre) for 2–3 days will contain approximately 30 mg a.i./kg. This is equal to the so-called 'no effect' level for Frescon in the food of rats, but still gives rise to concern about the widespread use of Frescon in areas where fishing provides the only source of protein.

In addition to work with molluscicides, COPR intends to carry out fish toxicity testing with a range of other pesticides that may contaminate tropical water bodies. In connection with the West African Onchocerciasis Control Programme (OCP), tests have begun which involve the organophosphate larvicide, Abate (temephos). The Pesticides Section of the Tropical Products Institute is co-operating in these tests by carrying out analyses of fish and water samples from laboratory toxicology experiments, and it is hoped to determine whether fish are able to concentrate Abate in their tissues. Studies of sublethal Abate toxicology are also in progress.

The laboratory work with Abate and other insecticides will be used as a back-up for fish pond trials which will be starting in Ghana in co-operation with the Ghanaian Environmental Protection Council. This project was initially conceived as being complementary to the OCP environmental monitoring scheme in West Africa, but its scope has been broadened to include other aquatic pesticides in addition to Abate. The interaction of these pesticides with the fauna and flora of small artificial ponds will be used to estimate the likely effects of large-scale vector control operations in the Volta River Basin. By training counterpart personnel, it is hoped that the testing facilities can ultimately be placed under Ghanaian control and form the basis for a national water pollution research centre.

Field work in The Gambia

In August 1975, P. H. Goll took up an appointment under Technical Assistance arrangements on secondment to the Government of the Republic of The Gambia, and is attached to the Medical Research Council Laboratories at Fajara. Part of his assignment is involvement in a project designed to more accurately estimate the longevity of the adult worm of *Schistosoma haematobium*. This will require an experimental interruption of transmission using molluscicides in an area containing a sample population large and stable enough to provide significant results at the end of about 4 years. The area selected is 200 miles inland on the Senegambian plateau where seasonal transmission of 2–3 month duration occurs in rainwater pools which appear in shallow depressions in outcropping cuirasse, and which support the intermediate host snail *Bulinus senegalensis* (Müller). Intensity of transmission is high enough to result in nearly 100% prevalence among teenagers of the villages associated with these pools. It is proposed that one series of pools associated with one group of villages will be treated with Bayluscide to prevent transmission completely, while the population of another group where no mollusciciding will be practised will be studied as a control; in the former group the rate of loss of worms will be measured by reduction in viable egg excretion.

So far preliminary field trials have shown successful elimination of snails can be achieved with an initial concentration of 0.3 ppm (a.i.) Bayluscide. A population census for the project area has been made indicating that the initial sample size will be in the region of 1100 persons in each of the "treated" and control areas.

Laterite pools are not the only site of transmission of vesical schistosomiasis in The Gambia. Some is associated with bolons (tributaries of the River Gambia) which vary from being seasonal to permanent, and support other species of snail host. It is also possible that to a lesser degree transmission may occur in rice paddies adjacent to the River. Irrigated rice is assuming increased importance as development schemes are established and this may create new habitats suitable for host species of snail hitherto unable to colonise in these areas. So another aspect of the assignment will be the continuing regular surveillance of rice and other irrigated crop developments throughout The Gambia for evidence of invasion by schistosome hosts.

2 CONTROL OF RIVER BLINDNESS IN WEST AFRICA

The task of the Onchoceriasis Control Programme (OCP) is to prevent the transmission of river blindness in the Volta Basin by controlling the larval stages of its vector the blackfly, *Simulium damnosum*. Reinvasion of the area by adult blackflies is possible as the vector is present outside the Programme Area. COPR was asked to study this problem in 1975 and in particular to try to assess the probability that major redistributions of *S. damnosum* are the result of windborne displacements.

Insects which are windborne move downward and their arrival in an area has been found to be associated with zones of convergence in the wind-field having passed over it. Dr J. I. Magor, Dr L. J. Rosenberg and Mr. D. E. Pedgley examined the weather preceding arrivals of *S. damnosum* to see if it was similar to that which has already been identified as being associated with arrivals of other windborne insects. These studies are still in progress but the first results show that zones of convergence, either the inter-tropical convergence zone or smaller zones within storms or both, had passed over the area. It is probable, therefore, that reinvasions of *S. damnosum* are the result of windborne displacement.

In considering future research on the migratory displacements of *S. damnosum*, the COPR team have suggested to WHO that the main objective should be to test the hypothesis that some or all of these flights are windborne and that an integral part of these studies should be to find out whether as a result of migration reinvasions of the OCP Area will be frequent or rare. So far, only circumstantial evidence suggests that windborne migration of *S. damnosum* occurs and so it has been suggested that direct evidence should be sought by trying to trap flies above their boundary layer. This should be done as part of a series of field-studies on flight behaviour around and above breeding sites in relation to changes in the level of population density, river state and weather. Priority should also be given to providing answers to questions which need to be answered before trajectories of reinvading flies can be constructed, i.e. laboratory studies to determine the air-speed, fall-speed and changing flight behaviour of *S. damnosum* in relation to age, physiological and nutritional status; the duration of flight and the stimuli to which the flies orientate during a flight. Biogeographical studies should be continued and should make full use of the data being collected by OCP by recording on maps the changing patterns in the distribution of *S. damnosum;* they should also assess the part played by control measures and natural changes in population levels in bringing about the recorded changes in distribution. By establishing where losses and gains occur concurrently in the OCP Area, a pattern of reinvasions for each area should emerge and over a number of years sufficient data should accumulate to allow forecasts of changes in the distribution of *S. damnosum* to be made.

3 TSETSE FLY CONTROL

Collaboration with the Department of Veterinary Services and Tsetse Control in Botswana was renewed by a visit in July by Mr C. W. Lee and Mr G. G. Pope. They assisted in airspray trials in the Okavango swamp area in the north of the country. The main objectives were to continue studies of the physical, chemical and biological performance of aerosols released from fixed-wing aircraft fitted with Micronair equipment. The experiments were carried out between spraying cycles of aerial operations being conducted by the Department to control *Glossina morsitans centralis* using a Cessna 310 aircraft.

Aerosols of approximately 36–61 μ vmd were produced from an AV 3000 Micronair unit fitted to the Cessna 310 and formulations of synergised pyrethrins in diesoline and endosulphan in very volatile and partially volatile solutions were evaluated. Synergised pyrethrins failed to give any significant reduction of *G. morsitans* at rates of 0.6 and 1.2 g/ha, although good reductions of fly numbers were achieved at about 6 g/ha with endosulphan in Shellsol AB. The more involatile solutions were very encouraging and it is planned to carry out further trials to assess these formulations in more detail.

G VERTEBRATE PESTS

1 BIRDS

Quelea investigations in Nigeria

In the third and final year of the project Dr P. Ward and Dr P. J. Jones have completed the programme concerned with the seasonal migrations performed by the Red-billed Quelea *Quelea quelea* and a number of lesser granivorous pest species. These include several species of doves *Streptopelia* spp., and starlings *Lamprotornis* spp., the Golden Sparrow *Passer luteus*, Yellow-fronted Canary *Serinus mozambicus*, and Village Weaver *Ploceus cucullatus*. The aim of this study was not simply to work out the nature of the population displacements, for it is important to know also what factors determine the timing of the movements, and the ultimate reason for migration. This information is essential to meaningful discussion of ways in which the damage to crops by various bird pests might be avoided. The main transect study, which formed the basis for the work, involved monthly visits to eight widely-separated stations along a roughly north-south axis between 13°30'N. near the north end of Lake Chad, and 7°N. close to the Cameroun border. This was completed in November 1974 after 14 months of continuous observations, but sampling continued on a weekly basis at the station close to Maiduguri until the end of the project in September 1975. In addition, information concerning the (mainly north-south) migrations obtained by the transect method was extended by four major expeditions during 1975. To the north of L. Chad the line was extended to the Massif du Termit, in Niger. In the mid dry season, and again in the rains, counts were made at points half-a-degree apart between 13°N and 15°30'N. Similarly, a southward extension was obtained by a dry-season and wet-season visit to a series of points from 70°N to 3°N, on the coast of Cameroun.

The results of the migration studies have furnished further proof of the departure of *Q. quelea* from the L. Chad region when the rains begin in July. They fly to the Benue Valley area in the south where they remain for only a few weeks before returning north to breed in August and September. An attempt to study the behaviour and diet of Queleas while they are in their southern quarters in July was largely thwarted by a scarcity of birds. Without doubt, the population crash suffered by the formerly huge Quelea population has been caused by the succession of droughts in the L. Chad region in recent years.

The Golden Sparrow has been shown to make regular migrations southward in the dry season from its breeding areas just north of the frontier between Nigeria and Niger. The dry season incursions of Golden Sparrow flocks now reach much further south into Nigeria than they did in former years — another manifestation of the drought. In contrast, the Village Weaver is a breeding visitor to the L. Chad region. Although a few seem to be resident in favourable localities, numbers are greatly swelled in the rains by the arrival of large flocks from the south of the country. They stay to breed returning south again soon after the beginning of the dry season. The movements of the Yellow-fronted Canary resemble those of *Quelea* (viz. south at the start of the rains and back north a few weeks later to begin breeding). Though not a major pest in most parts, these canaries cause serious local damage to early crops of bullrush millet in the south of North-East State where they congregate briefly in the course of the migration.

The migrations of doves are more complex, though the general patterns of movement have largely been elucidated by the transect technique. The study of Glossy

Starling migrations was not so successful, however. Though they certainly move about a great deal, as is shown by the systematic observations made along the transect, no clear patterns emerge. This is mainly due to the three species involved (*Lamprotornis chalybeus, L. chloropterus* and *L. chloronotus*), which have overlapping ranges, being virtually impossible to separate in the field.

Aside from these mainly observational studies, the team's main task has been the carcass analysis (for fat and protein content) of thousands of Queleas collected during the previous year and kept in cold storage. The results of the analyses are pertinent to two eco-physiological investigations carried out within the project period. The first was aimed at establishing the relationship between pre-migratory fat-deposition and length of migratory flight. The second sought to determine the physiological basis for breeding failure by Queleas in the sub-optimal habitats created by the recent desiccation of the Lake Chad region.

2 RODENTS AND SMALL MAMMALS

Rodent damage survey

As reported in 1974, the survey work for this project was completed and the further difficult task of collating the replies to the postal survey was finished during 1975 by Dr H. S. Hopf for COPR and at the Tropical Stored Products Centre of TPI. Publication of the final report is expected during 1976.

Rangeland management in Mexico

At the experimental farm "La Campana" in Chihuahua State in Northern Mexico, the government is carrying out research on how to improve the productivity of this semi-arid region, particularly in beef production. The work is supplemented by parallel experiments in the even more arid State of Sonora in the West. ODM agreed to provide technical assistance and Dr M. P. L. Fogden (of COPR) and his wife Dr P. Fogden are now in the third and final year of their investigations.

Dr Fogden is about to prepare a detailed report on their work but the following are their major conclusions as they stand at present. The animals concerned are the jack rabbit *Lepus californicus* (Lagomorpha) and several species of kangaroo rat *Dipodomys*. Conclusions are based on exclosure experiments at "La Campana" involving combinations of no grazing, moderate grazing and over-grazing on the one hand with natural rodent and lagomorph populations, exclusion of rabbits and exclusion of both rabbits and rats on the other – nine treatments in all. There has also been a survey of jack rabbits and their habitats over large portions of Northern Mexico.

In the short term control of these animals might produce more grass, but also more scrub and weeds. In the long term, control might not be beneficial in the absence of other measures, particularly scrub control. Killing alone would be quite useless. Better range management, i.e. brush control, reseeding, resting, possibly the use of fertilisers, as well as careful water conservation is the answer, and if these are effective the rodent problem will become negligible. Density of brush more than 50 cm high, percentage grass cover (the less grass, the more rabbits), diversity of weed species and the topography of the ground govern the density of the rabbit. Where scrub has been eliminated, rabbit populations have been reduced by 80%.

Of the kangaroo rats, two species are of minor importance and one could be important at high population density. They can be controlled by scattering bait on the mounds, but this is only feasible in open grassland as elsewhere mounds may be missed and costs be rather high. Most of the grass taken by these rodents is of little value to cattle. In fact, the kangaroo rat may be beneficial by keeping down undesirable plant species, but this has to be counterbalanced by the loss of space taken up by the mounds.

It is concluded that neither jack rabbits nor kangaroo rats are seriously limiting the production of beef. It is hoped to confirm this in the remaining season by analysing the faecal pellets for plant material eaten, comparing the diet of jack rabbits, cotton tails and cattle, and if this gives the results indicated by the work described above, control will certainly be uneconomical. In Sonora, where grass is very scarce and cattle browse on brush which also conserves the soil and can therefore not be eliminated, there is no possibility of eradicating the jack rabbit anyway. It might however be possible to utilise the jack rabbit by processing it as a protein supplement in animal feed.

H PEST CONTROL CHEMICALS AND APPLICATION METHODS

1 EVALUATION OF CHEMICALS FOR MOSQUITO CONTROL

The number of new chemicals submitted to the WHO Collaborative Scheme continues to diminish, but some of those examined recently are exceptionally potent as insecticides. They come from the group of newer, synthetic pyrethroids and are toxic to a wide variety of insects, including mosquitoes, as well as having other desirable properties. NRDC 143, for instance, has been mentioned previously as a compound with very prolonged residual activity. NRDC 161 is the most toxic insecticide to mosquitoes that is known at present, and it too is very persistent. It has shown no loss of activity over 32 weeks at low dosage rates from an emulsion spray, and it is now being formulated and tested as a water-dispersible powder on a range of surfaces which might be sprayed in the control of mosquitoes in houses.

Compounds from other classes have also been tested but show no advantages over existing insecticides. Attention is therefore being concentrated on the pyrethroids with a detailed examination of their biological and physico-chemical properties.

2 EVALUATION OF CHEMICALS FOR TSETSE FLY CONTROL

As described in the 1974 Annual Report new chemicals, formulations and methods of application are being sought as essential requirements for the projected large-scale tsetse fly control schemes. Candidate insecticides from different chemical classes have been evaluated for toxicity to teneral tsetse flies, *G. austeni*, by topical application, and again the only ones more effective than the organochlorine compounds currently in use, endosulphan and dieldrin, were synthetic pyrethroids. In tests carried out in the last year the most active organophosphate was crotoxyphos (LD50, 10 ng per fly), and the most active carbamate was bendiocarb (LD50, 12 ng per fly).

The most important event in the development of new chemicals in the year under review has been the commercial production of the pyrethroid NRDC 161 which has been mentioned already in the section on mosquito control. It has been shown to be extremely toxic to tsetse flies of various species, and it should be possible to use this high potency to reduce both dosage and concentration in sprays applied by both high and low volume techniques. This and other chemicals being considered as replacements in tsetse fly control have been formulated as water dispersible powders and solution concentrates and sprayed on to glass plates, plywood panels, ivy leaves and pieces of tree bark. Changes in toxicity of the deposits to flies and in the weight of active ingredient as determined by chemical analysis are being followed. Endosulphan is a very effective contact insecticide but lacks persistence, whereas tetrachlorvinphos and iodofenphos last much longer because they are less volatile but are less toxic by contact action. The pyrethroids NRDC 143 and NRDC 161, however, are both highly toxic and more persistent than the alternative compounds. Endosulphan, for example, has a half-life of 3 days in solution on leaves under the test conditions, and tetrachlorvinphos and iodofenphos a half-life of 15 days, while NRDC 143 suffers negligible loss in either toxicity or mass over 28 days.

3 EVALUATION OF CHEMICALS FOR LOCUST CONTROL

In addition to the evaluation of candidate compounds against *Schistocerca* and *Locusta* from the routine (crowded) cultures, the opportunity has been taken to carry out some tests on *Schistocerca* reared isolated from hatching (i.e. laboratory solitaria). These isolated insects, as they become available on completion of another experiment, are considerably older than those used in routine evaluations, and a parallel series of tests is therefore being made with crowded *Schistocerca* of comparable age. In the isolated locusts, up to about 5 months after fledging, no differences in susceptibility have been found from the crowded insects normally tested 2—3 weeks after fledging. In the case of the synthetic pyrethroid permethrin, however, the crowded insects are considerably more susceptible at 3 months than at 2—3 weeks.

4 BIRD CONTROL — AVICIDES

Limited testing of candidate avicides was carried out in the latter half of the year and one avicide that appears promising is undergoing further tests.

5 MISCELLANEOUS CHEMICAL WORK

A substantial proportion of the time is necessarily devoted to the development and use of chemical analytical methods for newer and experimental insecticides. The compounds which are currently of particular interest are the synthetic pyrethroids and quantitative chromatographic methods are now available for the most important of these. Such methods are being employed in a study of those physical properties of the best new chemicals which are important in controlling their persistence in spray residues and the environment. These include evaporation from various formulations on different surfaces and decomposition in tropical soils.

Another big effort during the year has been concerned with the examination of formulations used for tsetse fly control. This has required a survey of commercially-available solutions which are supplied for ultra-low volume techniques and includes determination of content of active ingredient, separation and identification of the solvent or solvents and measurements of their volatility, viscosity and solubility powers for insecticides. New compounds have been formulated as solutions, emulsions and dispersions and sprayed in a tower which has been modified recently for use at very low as well as high volume rates.

A variety of samples submitted by other organisations both at home and overseas has been examined. These have included lindane dusts and DDT emulsion concentrates from Nigeria; pesticide residues in food for rabbits and goats used for tsetse fly breeding in Tanzania; DDT and carbaryl in soils and vegetation from Kenya; insecticides and oils on test papers used in bioassays with ticks and oils and surfactants in emulsion concentrates.

6 AERIAL SPRAYING OF COTTON IN SWAZILAND

Some failures of the ultra-low volume aerial spraying of cotton in Swaziland during the 1973—74 season were discussed with the Cotton Research Corporation entomologist Mr N. Morton, and proposals were made for a short study of air spray problems by a COPR team during the 1974—75 season. Subsequently, at the request of the Swaziland Ministry of Agriculture and the Swaziland Cotton Board, Mr D. R. Johnstone and Mrs K. A. Johnstone visited the Lowveld Experimental Station in

Fig. 24. Calibration of a spray plane at the Big Bend Research Station airstrip in Swaziland.

Fig. 25. A Micronair AV 3000 atomiser fitted on a Cessna 188 Ag-Wagon aircraft at Big Bend in Swaziland for pesticide trails.

February/March to determine the limits of meteorological conditions outside which satisfactory spray deposition might not occur in cotton when oil-based and water/molasses-based sprays were applied from the air at ultra-low volume (up to 5 l/ha) and very low volume (10 to 15 l/ha) rates.

Field and laboratory work were conducted at the University of Botswana, Lesotho and Swaziland Lowveld Experiment Station at Big Bend with aircraft calibration and assessment trials at the station airstrip and on a neighbouring cotton estate. Owing to the exceptionally wet season it was not possible to study deposition in the field under the conditions normally prevailing in the Lowveld during January to March when high temperatures and low humidities can call for restrictions on aerial spraying. It was therefore necessary to assess the implications of laboratory test work on the volatility of formulations against the background of field and airstrip trials made under less stringent conditions and to draw some comparisons with experience in other tropical areas faced with similar spraying problems. Amendments to the existing recommendations were proposed to define formulation and droplet size requirements more precisely as well as the meteorological conditions necessary for effective application of ultra-low and very low volume sprays.

7 SPRAY TRIALS ON COTTON IN THE GAMBIA

Mr W. J. King visited The Gambia for the duration of the cotton season from July to December at the request of the Director of Agriculture to advise on cotton spray trials. The Cotton Development Project is confined at present to the Upper River Division some 200 miles up-river from the capital Banjul. The major insect pests of the crop include the American, the red and the spiny bollworms and the whitefly *Bemisia tabaci*, and their control is essential if profitable yields are to be produced. Trials were conducted to compare different spray methods and regimes. Results indicated that the very low volume applications of wettable powders suspended in water were as effective as conventional knapsack sprays and ultra-low volume applications of oil-based formulations and were appreciably less costly.

A 100 hectare pilot scheme to evaluate the response of the farming community to the introduction of new equipment and techniques is planned as part of the programme of cotton expansion implemented by the Department of Agriculture. This will involve the field training of extension staff not yet familiar with the new application methods since they will ultimately be responsible for the planned transition from the knapsack sprayers used at present to the ULV sprayers. The training programme was initiated, therefore, with demonstrations and lectures to members of the Crop Protection Unit of the Department of Agriculture.

8 PESTICIDE APPLICATION TO CROPS IN THAILAND

Research on the application of pesticides in Thailand is carried out by the Pesticides Application Section of the Division of Entomology, Department of Agriculture. In response to a request from the Thai Government for British technical cooperation in this field, ODM initiated in June 1974 a Pesticides Application Research Project staffed by Dr H. Rendell and Mr J. Sutherland. During October Mr D. R. Johnstone made a short visit to Thailand to establish technical liaison with the two groups, particularly with regard to cotton insect control.

Pesticides in Thailand are most widely used on cotton and vegetables. Field trials carried out at the Cotton Development Centre, Tafka, in the Central Region this season have been aimed at comparing alternative formulations of the insecticide triazophos as well as ULV and high volume applications. During the visit to Tafka some preliminary field work was carried out to develop a technique for estimating the total spray intercepted by individual cotton plants in order to assess the extent to which formulation could affect spray recoveries under different meteorological conditions.

9 TESTS OF ULTRA-LOW VOLUME SPRAYERS

Droplet performance characteristics of a prototype ULV rotary atomiser intended for aerial spray application and of a new high performance hand-held rotary atomiser have been assessed in the laboratory at Porton. The measurements have suggested improvements to the liquid feed system of the former. The latter machine offers the possibility of producing considerably smaller spray droplets, of less than 50 μm volume median diameter, than its forerunners and is of special interest for ultra-low volume spray application to field crops. The behaviour of such small droplets is largely dependent on the prevailing micrometeorological conditions, and an extended study of the field performance of the machine and of factors which may promote or limit its effective use is desirable. This investigation has been started.

10 EVALUATION OF THE EVERS AND WALL MK II EXHAUST NOZZLE SPRAYER

The exhaust nozzle sprayer is a simple machine operated by utilizing motor vehicle exhaust gases. It was originally designed and developed for the ultra-low volume application of a specially formulated solution of dieldrin in oil to vegetation for locust hopper control.

The first models of the sprayer were for use with a petrol-engined Land Rover, but in recent years a variety of vehicles with different engine capacities has become available. The Mk II model of the sprayer has been produced by the manufacturers, Evers and Wall Limited, Lambourn Woodland, Newbury, Berks., with modifications, to enable it to be used on most vehicles with engine capacities of 2–3.5 litres. Briefly the main modifications are a flexible heat-resistant rubber hose for feeding the exhaust gases to two 50-litre insecticide tanks which can be used separately if required for two different chemical solutions. A single spray stack with detachable feed tubes and gas restrictors are supplied, which are designed for use with vehicles of different engine capacities. Other modifications have been made to tank-filling/sealing covers, drain valves, fillers and on/off controls.

The evaluation of the sprayer has proceeded in conjunction with the Overseas Spray Machinery Centre, Silwood Park, Ascot, Berks. Tests have been carried out using a standard petrol-engine Land Rover and a diesel tractor (3.5 litre engine capacity) and have included estimates of gas throughputs at various engine speeds, the effect of back pressure and engine revolution speed on throughput of spray liquid from the liquid feed tubes, exhaust spray stock temperatures and some droplet spectrum analysis.

The results of the tests of liquid throughput and droplet spectra produced by the Land Rover system were satisfactory and compared favourably with older models.

For back pressures of 0.2—0.38 Bar the outputs were 950—1200 ml/minute. With the diesel tractor the liquid throughput with the larger feed pipes was excessive, ranging from 1760—4700 ml/minute. The latter tests also produced larger drops with a volume median diameter of 92 μm.

The use of exhaust gas in the spray system results in the rapid heating of the stack pipe and enclosed liquid feed pipe. By using a thermocouple potentiometer with iron-constantan probe, the temperatures within the spray system were measured. Within the stack pipe the temperature rose to 90°C—120°C after five minutes, and 87—95°C in the liquid feed pipe. The temperature of gas 10—15 cm above the emission point fell to 42—46°C when the ambient temperatures were 9—13°C.

Further tests are to be carried out with the Mk II model, to modify the spray stack system and utilise a single feed pipe and restrictor system for use with most vehicles.

11 HOVERCRAFT

Further performance tests were carried out in the United Kingdom and as these were satisfactory the craft was sent to Botswana for further trials under tropical conditions. The trials were successful, the craft coping well with weedy conditions in the swamps of the Okavango Delta, and a Mk II craft with improved rudders and larger engines has been ordered. This should be a considerable improvement on the earlier model, and it is hoped to carry out further overseas trials in the coming year.

12 METHODS FOR THE DETECTION OF ACARICIDE RESISTANCE IN TICKS

In conjunction with FAO's plan for a Global Acaricide Resistance Monitoring Programme, COPR agreed to assist in the development of a test kit by making independent assessments of methods for the detection of acaricide resistance in ticks.

A method used routinely by CSIRO in Australia, using packets made up from filter paper impregnated with olive oil solutions of acaricides, has been studied mainly with larvae of two strains of a single-host cattle tick, (*Boophilus microplus*), a fully susceptible (Yeerongpilly) strain and an OP resistant (Biarra) strain.

Possible alternatives to olive oil as the non-volatile solvent were tried in an attempt to overcome objections to olive oil because of its undefined nature, its liability to chemical oxidation and biological degradation and its poor solvent properties with some acaricides. Control papers impregnated with polyethylene glycol 400, Risella oil, di-octyl phthalate (and some related esters) or light liquid paraffin all caused appreciable toxicity to tick larvae enclosed in packets for 24 hours. Reducing to a quarter the amounts of some of the solvents on the papers reduced but did not eliminate mortalities. Heavy liquid paraffin proved to be non-toxic to larvae of *B. microplus* after several days exposure, a feature essential for any possible substitute for olive oil. Acaricides were generally more effective in liquid paraffin than in olive oil, LC50's being lower with the former by factors of 2 to 4, although resistance ratios between the LC50 values with resistant and susceptible strains of *B. microplus* were similar with both solvents.

Work is continuing on the packet test with investigations of the storage life of impregnated papers, the solubility of acaricides in olive oil and liquid paraffin and the effects of variations in the test conditions (e.g. humidity and temperature).

Assessments are also being made of a larval immersion test method widely used both for the screening of potential new acaricides and for resistance monitoring.

13 NRACC SUB-COMMITTEE ON PESTICIDE APPLICATION OVERSEAS

This Sub-committee (Chairman, Dr P. T. Haskell, Secretary, Mr C. W. Lee) held two meetings during the year, one at TSPC, Slough and the other at NIAE, Silsoe.

The Sub-committee is made up of representatives of six scientific organisations concerned with pesticide application research overseas. The main objectives of the Sub-committee are to promote and co-ordinate this research as requested by overseas countries, to liaise with UK organisations and to provide the NRACC with advice and assistance on pesticide application.

The work of the Sub-committee during the year has been concerned with a wide variety of problems as, for example, the control of premature fruit drop of citrus in the West Indies. A project has now been implemented with the appointment of a Plant Pathologist in Belize. The Sub-committee has advised on a programme of research in the development of zero/minimum tillage techniques on peasant farms in Nigeria through the IITA in Ibadan, in particular on herbicide application for weed control, ULV techniques and the suitability of a hand planter capable of planting seed directly into stubble. Evaluation of prototype equipment for ground and aerial dispersal of pesticides has been coordinated by the Sub-committee and the work undertaken by both OSMC and COPR.

I ENVIRONMENTAL STUDIES OF PESTICIDES

Introduction

Two previous reports have been produced on this work (see ODM/IITA Pesticide Residues Research Project 1st and 2nd Interim Reports, 1974 and 1975) in which details of experimental design and methodology are described. An initial series of publications dealing with major aspects of the project is now in preparation. The work has been conducted at the International Institute of Tropical Agriculture, Ibadan, Nigeria, with support from ODM Research Scheme R2730. This funding arrangement ceased on March 31st 1976 and certain elements of the programme will be continued under the auspices of the IITA Farming Systems Programme for a further two years.

Experimental design

The project has been designed to study the fate and biological significance of DDT in soil resulting from its application to cowpea, *Vigna unguiculata*, for the control of leaf-feeding pests. Particular attention has been given to potential effects on soil fertility. Field plots have been established with the following treatments in a 4 x 3 randomised block design.

1. *Bush* — maintained as regenerating secondary forest
2. Cleared, cultivated and *untreated* with pesticide
3. Cleared, cultivated and *treated* with pesticide in consecutive growing seasons
4. Cleared, cultivated and treated with pesticide to form a *contaminated* soil treatment during unsprayed seasons.

In the first growing season of 1975, season 75/1, sub-plots 2m x 12m at the end of each cultivated plot were treated with maize straw, incorporated to a depth of approximately 20 cm and applied at a rate of 24.7 t/ha, to investigate the response of different treatments to supplementation of soil organic matter. DDT was applied weekly from planting to harvest as a 0.25% water-borne emulsion to give an even application of 1 kg a.i./ha on each occasion. Ten such applications were made in each growing season. In seasons 75/1 sub-plots of the untreated plots were sprayed to compare the performance of protected plants in all three treatments.

In season 75/1 there were signs of serious pathogen buildup, probably associated with continued sole cropping of cowpea. There was a heavy infestation of leaf spot, associated with the fungus *Protomycopsis phaseoli*, and other pathogens such as *Cercospora cruenta* were also present at high levels. For this reason the cropping pattern was changed in season 75/2 and soybean was intercropped with the cowpea to avoid total crop failure should cowpea pathogens continue to increase. It was also felt that the soybean could serve as an additional indicator of soil performance in the three treatments. Over the three years of the project studies have been undertaken in four major areas. There are considered separately below.

Pesticide studies

Investigations of the fate and distribution of DDT and its major breakdown product, ppDDE, have been undertaken with a view to constructing a model of its physical and chemical behaviour within the crop ecosystem. Such information also provides a background for the interpretation of biological phenomena observed.

Work in 1974 based on measurements of crop cover indicated that over the whole spraying season approximately 70% of the pesticide applied caused direct soil contamination. This was verified in season 75/1 using a method in which a dye, Croscene Scarlet, was incorporated into the spray emulsion for one half-plot on each spraying occasion. Immediately after spraying, individual plants were removed and aqueous washings from them analysed spectrophotometrically to assess the amount of dye and therefore spray striking each plant. A series of glass fibre discs arranged over the plot in different positions relative to the plant cover were treated similarly in order to ascertain the amount of dye reaching the soil directly.

The amount of pesticide in the soil over the 0–300 mm horizon appears to be increasing by relatively smaller increments in the later additions of pesticide than in earlier ones. In general, spraying results in an increase in the relative concentration of DDT in the 0–50 mm horizon with redistribution mainly occurring during ploughing. This has again been demonstrated during 1975. An interesting feature of the results is the overall loss of pesticide from treated plots during the second growing season, despite the addition of DDT at the normal rate. This may be associated with a very much greater plant cover during 75/2 due to the presence of soybean, which would have certainly reduced soil contamination considerably. Studies on the rate of loss of pesticide from sprayed leaves and soil suggest that up to 50% of the amount applied may be lost within one week and air entrainment within the plots has shown there to be a considerable increase in the amount of DDT present in the air following spraying. Comparison of the relative residue levels in the various organs of plants growing in contaminated soil has shown there to be a number of discrepancies between those plants that had received a foliar application and those that had not, particularly in relation to what might be expected from the application of a non-systemic insecticide. These results suggest that there may be movement of DDT from the highly contaminated leaves, in the case of the sprayed plants, into the roots. Evidence has also been found to suggest that some of the DDT absorbed by the roots from contaminated soil may be further translocated into aerial parts of the plant. The possibility of translocation of DDT within the plant from areas of high residue level to ones of a lower level will be reinvestigated in the coming year.

Population studies

The majority of population sampling in 1975 was confined to a one month period at the end of the first cropping season, in which estimates of subterranean and surface active fauna were made using the techniques described in previous reports. More satisfactory quantification of surface active fauna was achieved using a quadrat method in which population estimates were made at different times of day to allow for any bias due to diurnal variations in activity of the various groups sampled. The trend has been towards the establishment of species better adapted to the cultivated habitat after an initial fall in numbers due to disturbance. In 1973 there were few indications of pesticide effects on the subterranean fauna though some surface active groups more vulnerable to direct mortality during spraying were caught in smaller numbers in the treated plots. This was particularly the case for spiders. Lycosids (Wolf spiders) form the predominant element of the spider fauna, probably playing an important role as predators, and their populations have been consistently depressed in treated plots over the three years. Other predators affected have been carabid beetles, especially the smaller species. Ants dominate the surface active fauna in numbers, biomass and species diversity, performing a wide variety of ecological functions. Pitfall trapping gave higher catches throughout the period in treated plots and changes in species dominance were apparent, e.g. one species of *Pheidole* accounted for 37% of the total ants trapped in untreated plots in 1975 whilst representing only 2% of the catch from treated plots. Quadrat samples for absolute population estimates showed significantly lower numbers of ants in treated plots during season 75/2. Changes in the litter feeding surface fauna, principally gryllids and millipedes, remain difficult to quantify because of their low population density. It will be important to accomplish this since feeding studies have shown that they may be responsible for considerable organic matter turnover.

Dominant members of the subterranean fauna are the microarthropods, mainly Collembola and mites, earthworms and termites. The relatively shallow soils of the site do not support large populations of termites and the scale of the plots is small in relation to termite foraging behaviour, such that direct population estimates are difficult. However, a softwood baiting experiment showed no attack in treated plots with levels in untreated plots comparable to those found in bush. Earthworm behaviour appears to be affected by DDT. The activity of predominant species has been monitored by determining casting rates and the response to cultivation has involved an apparent replacement of *Hyperiodrilus* spp. by *Eudrilus* spp. Pesticide has reduced cast production by the latter species to a very low level in treated plots in parallel with increased activity in untreated plots. Direct sampling, however, shows no difference between cultivated treatments in biomass or numbers except for a possible trend towards smaller individuals in the treated plots. Further comment must await detailed species analysis of the samples. Numbers of microarthropods in treated and contaminated plots have fallen in comparison to those in untreated plots. Sampling in 1975 showed populations of oribatid and prostigmatid mites to be approximately halved in contaminated plots relative to those which remained untreated and even further reduced in treated plots. The most affected group was the predatory Mesostigmata, down to 2.5% of the level in untreated plots. These animals are probably important in regulating collembolan populations and a corresponding increase in the proportion of Collembola has been observed.

Decomposition studies

Decomposition of organic material has been studied using a wide variety of substrates presented in nylon bags of different mesh sizes. The bags have been placed in the field either at or just below the soil surface and recovered at intervals to assess weight loss of the substrate. Experiments with cowpea litter have shown a consistent, though not always significant, depression in decomposition rate when measured in this way associated with both contamination of soil and substrate. This effect has been most marked with buried stem and root tissue, where a slower decay tends to reveal differences in rate more clearly, and in series involving coarse (10 mm) mesh bags. Where fine (125 μm) mesh bags were used differences were small, suggesting that the microbial component of decomposition is little affected by DDT, an observation that is supported in general by data from population studies. In the coarse mesh bags, where the fauna is able to penetrate freely, reduced populations are associated with a lowered decomposition rate. A repeat in 1975 of the experiment conducted in 1974 (see report for that year) using DDT treated and untreated litter showed a greater reduction in decomposition rate for buried stems. In untreated soil with treated litter in coarse mesh bags the degree of the decomposition rate increased from 25% (ns) in 1974 to 44% (p.0.05) in 1975. Trials carried out with artificial substrates showed no significant differences associated with different degrees of contamination in Grid Plot (see previous report) soils. This may have been due to the homogeneity of the substrate reducing the importance of population diversity in the decomposition process. Experiments similar to those with cowpea were undertaken using cellophane presented buried and on the surface in medium (1.5 mm) and fine mesh bags though data from these is not yet available.

Some investigation of the factors operating in decomposition was started in 1975. Work on a common species of surface feeding oribatid mite showed it to be capable of consuming up to 140 kg/ha/annum of cowpea leaf material at the high population densities sometimes encountered. Consumption studies with gryllids indicated that at population levels recorded from quadrat sampling, almost certainly an underestimate, they could account for 110 kg/ha/annum. Millipedes probably play a much smaller role. Studies on the interaction of fungivorous microfauna and fungal invasion and colonisation of material are in progress to investigate the dynamics of this system and its role in decomposition. Increased grazing may stimulate breakdown by promoting fungal growth and there appears to be a relationship between microarthropod population, hyphal turnover and cellulose decomposition.

Crop studies

The decline in crop yield commonly associated with repeated cultivation has continued into 1975, though with a surprising increase in yield in the untreated plots in 75/1 such that no significant differences occurred between treatments. This was probably associated with a low level of pest attack with the major damage attributable to plant pathogens such as *Protomycopsis* and *Cercospora* which would be little affected by the pesticide treatment.

In contaminated plots yield has followed that of treated plots for seasons in which the former were sprayed, but with a tendency to be higher. For seasons where these plots were unsprayed these yields were close to those from untreated plots with a tendency to be lower. This would be the expected situation if soil contamination was indeed having an effect on soil productivity. If log yields are compared for unsprayed seasons in untreated and contaminated plots the rate of decline in the latter is significantly ($p > 0.05$) greater. During season 75/1 sub plots within the untreated plots were sprayed with DDT to compare their performance with that of treated plots. Relative initial productivities of the two treatments can be calculated from the data of season 73/1, in which neither was sprayed. Treated plots gave 1.43 times higher yield. Comparison in 75/1 with both sprayed gives a factor of 0.90, a further indication of a relatively greater decline in situations where DDT has been incorporated into the soil. Caution must be exercised in interpreting these figures since considerable clarification of the nutrient situation is still necessary. The removal of larger amounts of specific nutrients required for seed production from treated plots by virtue of the higher yields may have caused depletion, resulting in inhibition of seed production and this must be investigated. Chemical analysis conducted to date shows no significant differences in major nutrients though it is conceivable that a trace element may be involved.

Yields from the soybean planted in 75/2 as an intercrop with cowpea are difficult to interpret since insufficient attention was given to plant density effects, though estimates based on individual plant yields gave the highest values in untreated plots.

V INFORMATION

A SCIENTIFIC INFORMATION AND LIBRARY SERVICE

A substantial all round increase occurred in the use of SILS services by staff and external users in 1975 compared with 1974, as shown below, and thus continued the upward trend observed ever since the formation of COPR.

Following last year's internal talk, opportunities were made to inform COPR staff in the outstations and those on home leave of the new services offered by SILS, including computerised retrospective searching of various data bases which began last year. This communication has probably contributed significantly to the greater use made of SILS services.

During the year special attention was given to further SILS staff training in the use of computerised methods of information retrieval. On the other hand, the use and performance of indexes previously prepared manually in house were studied and it appeared that they were fulfilling a real need. Constant reference was also made to the wide range of abstract journals held, with their comprehensive indexes.

The Open Days provided an opportunity to welcome a large number of visitors and eight specially prepared display panels enabled SILS staff to introduce, demonstrate and discuss their work on these occasions, as well as later on in the year with other visitors including parties of students.

For some years past a review of SILS has been considered and the Deputy Director, Mr T. Jones, began the review of SILS, together with other parts of the Office of the Director, in February. In addition, in March, the Centre was fortunate in securing the services of its advisers in information and library matters, Mr M. J. Rowlands and Dr H. Neville to carry out an independent survey. The recommendations of the first review were made available to staff together with the Advisers' report in October. They concerned mainly structure and the re-allocation of certain kinds of work within the public relations, publicity, training, photographic and publications sections. The most crucial finding, however, was that there was a need for additional staff in SILS. Action to implement the recommendations is awaited in 1976.

LIBRARY SERVICES AND STATISTICS ON USAGE

It is quite evident from the statistics for 1975 that the essential services are being used more and more by our staff and other clients and that the work load is still increasing in all major areas. The percentage increases over 1974 are as follows:

Loans to outsiders (books and reprints)	up	82%
Periodical loans and circulation	up	50%
Photographic loans (prints and slides)	up	50%
Internal loans (books and reprints)	up	35%
External borrowing and photocopies	up	16%
Major information enquiries	up	15%
Accessioned items	up	9%
Xeroxing	up	2%

In addition, reports cataloguing, intake of annual reports, minor requests for information and abstracting have remained roughly at or just below the 1974 level. Just as

1975 showed an increase over 1974, it showed a very substantial increase over 1971 when COPR was formed. The 1975 figures are, in fact, in several cases, three, four and even five times the level of 1971.

Main library

Several new measures for greater efficiency were introduced during the year, from the redesigned reprint order cards (Mr P. Shannon) and internal forms for book purchase requests by Miss Lumley to the introduction of short term loans by Miss Sherwood when a new used book is wanted by a number of readers.

The concept of uncatalogued resource material was brought in by Miss Wortley to take care of promising material held in taxonomic or other subject order, to include material received as gifts but not of immediate interest to the library for incorporation, but which has been proved to be most valuable at times in answering unusual questions, or when new areas of work were being embarked upon.

Items accessioned into the library numbered 4960. Of necessity, a shorter way to prepare the master copy of the Library accession list was tried by Miss F. Sherwood, resulting in a 20% reduction in time taken. It remains to be seen whether, in the long run, if more staff time becomes available, it may be preferable to go back to the previous method as it gives better control of stock. The periodicity of the pest accession lists was changed to even out and ease production. Attention was paid to building up and curating stock in the outstations, particularly literature on molluscs. Weeding of duplicates, out of date and unwanted material continued as time permitted.

The classification scheme was overhauled by Miss P. Schofield to include some new subject areas like economics and to divide up some categories like invertebrate pests. More remains to be done and will link up with work on indexing when time permits.

Internal loans to staff compared with loans to outsiders are in the approximate ratio of 8:1. This year loans to staff numbered about 2500 while loans to outsiders were 335. These figures were the highest ever. The needs of the staff are also met by recourse to the excellent and vital services of the British Lending Library both for further loans (551) and for photocopies (699). The level of demand for this service seems now to have stabilised. Request forms are always checked prior to despatch to ensure that the items are not already available within the COPR system. The loans are almost exclusively books and monographs.

Special efforts were made to service the needs of staff overseas especially in Thailand, Upper Volta and the West Indies, including meeting their requests for photocopies via the British Library. Back-up literature services for projects abroad have been much appreciated by the recipients. As in previous years, we have also depended on academic and learned Societies' libraries locally for many items.

On the basis of observations of library use two future developments are envisaged, one, a geographical library for information on countries as well as their climates, soils, fauna, floras, etc., the other, to collect the abstract journals together in one room for ease of access for all users and to be nearer to those who most frequently consult them, namely the information scientists.

Periodicals library

A further 400 periodical titles have been added towards the full list of periodical holdings which now number more than 1200, six times as many as in 1970, largely

Fig. 26.
Miss P. Wortley, Head of SILS, assisting a visitor to the Periodicals Library.

Fig. 27. Some of COPR'S work on locusts being demonstrated to visiting scientists from Russia in August 1975.

as a result of the merger of foreign journals received in exchange for *PANS*.
Mr K. Smith enlarged the recording of holdings in the Kardex system by 69 titles.
A system of recording journals on receipt was also introduced in *PANS*. As in the main library emphasis was put on back up services for abroad, endeavouring to follow through the exercise begun last year, contents pages of journals were sent regularly on request to staff overseas.

An attempt was made to fulfil the needs of the new agricultural economist for journals not held by the Centre by restarting the practice of requesting circulation of relevant journals from ODM library and at the same time bringing in a category for economics in the main library.

For forward planning an analysis was made of the returned loans to check which subject areas of the library had been the most used. Predictably, the locust library, which is world renowned, is still the most heavily used by outsiders, as well as being almost eight times more heavily used than any other subject. The top twenty subjects were in order, following locusts, control (161), physiology (149), ecology (65), stemborers (61), meteorology (60), armyworm (59), agriculture (54), entomology (51), insect vectors (47), reference (36), birds (31), statistics (30), biology (28), pesticide residues (25), population (23), East Africa (22), Asia (21), scientific techniques (21), library science (17) and cotton pests (16). It is not surprising that the second most heavily used part is the control section, but, what is interesting is that four times as many books on the subject were borrowed as against reprints in this section, whereas in the locust collection reprints are borrowed eight times as often as books.

A particular feature of the use of COPR stock was well illustrated this year in that there are sudden surges in demand for some subject areas, as well as for entirely new subjects. The interest this year in literature on stem borers and on East Africa and Asia directly reflected the staff's new research interests and theatres of operation. The use pattern can change very quickly. The resource material, mentioned above, together with a comprehensive selection of basic literature on all kinds of pests have proved to be valuable in these circumstances.

Loans of periodicals this year are estimated to have numbered 776 and loans on regular circulation 852, an increase on previous years.

Weeding

As a result of the amalgamation of largely periodical literature received in exchange for *PANS* with that already held in COPR, a number of duplicates were isolated, together with out of date items, for disposal. The process, which began towards the end of 1974, of offering these items through the British National Book Centre was continued on a larger scale in 1975 by Miss J. Ufton. As a result well over 1000 parts were sent on request, many to libraries in developing countries, namely, Argentina, Chile, Egypt, Fiji, Malawi, Malaysia, Sierra Leone, Tanzania, Tunisia, Zaire and Zambia. European libraries in Belgium, Czechoslovakia, Denmark, Finland, France, Holland, Italy, Poland, Romania and the USSR also benefited.

Reports Library

The number of reports accessioned in the year was 474, and the new filing cabinets were installed to contain reports while the number of annual reports received and accessioned was 87. Fiftyfour new titles have been added since the first list of annual reports was compiled in 1974, making a total now of 314. Growing use is being made of this collection and the importance attached to the curation of this material has been amply justified.

Visitors

The overall number of visitors at 113 was 6 down on 1974. Twentyfive visitors came directly from less developed countries, a similar number from British government departments and a larger number from British universities, which included, however, students from overseas. Visitors from Europe numbered two (France and Norway), from South America (Guyana) one, from North America (USA) three, from Asia (India, Indonesia, Pakistan, Taiwan and Thailand) eight and from Africa (Kenya, Libya, Mali, Nigeria, Senegal, Sudan and Tanzania) sixteen.

Trainees

During the year students on pest control courses from Imperial College, as well as those studying librarianship and information science e.g. from Leeds Polytechnic, visited SILS. The displays prepared for Open Day were well used during the year for instruction purposes.

Two visitors who came individually to study SILS methods were Mr Hernandondo from Indonesia and Mr A. S. K. Atsu from Ghana, whose three week stay was arranged in conjunction with the British Council.

ABSTRACTING AND INDEXING

Pressure was maintained throughout the year to keep up the flow of published abstracts. Mrs J. Ridout continued with *Acridological Abstracts* new series and parts 5—8 were published in the four issues of *Acrida* (663 abstracts plus author index). The subject index to parts 1—4 was ready for the printers and that to parts 5—8 was in preparation. It was agreed that this should be published by the Centre itself. Two numbers of the original series of *Acridological Abstracts* (1975 nos 1 and 2) were also published (200 abstracts). The overall throughput could not have been achieved without the assistance of outside abstractors and indexers Mrs J. Stanley, Mrs D. Brinklow, Mr M. Dennis, and Dr R. Blackman, as well as Miss I. Tan, vacation student.

Further thought was given to the production of a bibliography on termites with abstracts in collaboration with Dr E. Ernist of the Swiss Tropical Institute and Dr R. L. Araujo of the Musea de Zoologia, San Paulo, Brazil. In this connection Mr P. Shannon visited the MRC Project Fair with its novel indexing system.

Indexing has occupied a considerable amount of SILS time this year and has resulted in the maintenance or introduction of the following:

1. *COPR weekly bulletin*   annual index to contents   Miss P. Schofield
2. *COPR publications*   card index to subjects   Miss P. Schofield
3. *PANS* Cumulative Index   Vols 1—18. 1955—1972   { Miss J. Broughton
                                                        { Miss D. Brinklow
4. *Acridological Abstracts*   subject index   { Miss J. Ridout
                                                 { Mrs J. Stanley
5. Pests/crops/countries index   { Miss P. Wortley
                                      { Mr A. Green

6. Library (subjects excluding locusts) index to contents   Miss F. Sherwood
7. Trade literature index   { Miss P. Wortley
                              Dr M. N. D. B. Yeates
8. Mollusc literature   author and subject index   { Miss P. Schofield
                                                     Miss F. Sherwood
                                                     Mr P. Shannon

The index to the library contents is incomplete due to lack of staff time and pending a joint decision being made on indexing by all three Scientific Units of the ODM, working together on the SIFT Committee. A small pest/crops/countries card index to selected existing literature in COPR library was produced but has not been added to for lack of time, though it has proved a useful first point of entry to the literature in several searches.

Though literature of this type can be very successfully retrieved, except perhaps for countries, by means of a wide range of existing printed subject indexes to abstracts; it is then necessary to find the location of the item in the library by reference to the author catalogue. To eliminate this last step the compilation of a retrospective index based on the library catalogues has been considered — an immense manual task to which there appeared to be no short cut. Such a compilation would also be valuable in weeding the library to create more space. For these two reasons the daunting task was considered in several ways and finally embarked upon. By the end of the year with help from vacation students a start had been made.

Individual scientists have literature collections and often find these become too bulky and soon out of control. In the autumn Miss P. Schofield, Miss F. Sherwood and Mr P. Shannon embarked upon sorting of the mollusc literature on request for the Molluscicide Unit at Grays Inn Road. An author card index was prepared in which the research staff collaborated and was partly typed by the end of the year. From these cards, copies will be made to compile a subject index.

INFORMATION RETRIEVAL AND INFORMATION ENQUIRIES

A wide range of literature has been retrieved by extensive use of abstract journals by information scientists on the premises, as well as by a number of retrospective searches of computerised data bases like BIOSIS, CAIN, and UKCIS on line and off line. A number of searches concerned subjects new to the Centre such as date palm scale and mites, sudden death in cloves and citrus greening disease, while others were the direct outcome of COPR projects such as cyanogenesis in cassava and aerial spraying of tsetse flies. A bibliography on the safe handling of pesticides (Miss F. Sherwood) was one example arising from the liaison of the Centre's staff with workers elsewhere in the world. Other searches arose from the needs of COPR staff abroad. Information is being collected on existing relevant computerised systems and their availability. Opportunities have been taken to use as many as possible to gain experience.

For information retrieval in answer to enquiries from outside, manual methods have also played a major part. The new card index to COPR publications has proved extremely useful. The PANS Cumulative Index is available in printed form for those who wish to have it beside them, but opportunity was taken after its compilation to re-arrange the index cards under specific headings as distinct from the purely alphabetical arrangement in the printed version. This has been found to be a valuable retrieval tool in itself for a wide range of queries.

Information may be retrieved on locusts and grasshoppers now through a continuing index albeit regrettably in two forms. There is a card index in two parts, one for subjects and the other for species, as well as printed indexes to *Acridological Abstracts* old and new series arranged alphabetically and hierarchically, including species in with the subjects in one sequence. This tool has been in constant use by COPR staff and visitors who consult it directly, as well as by COPR's information scientists answering enquiries.

An analysis has been made of information enquiries received in 1974 and 1975 by (1) their countries of origin, (2) type of originating organisation or individual and (3) their subjects. The queries have been grouped in the first case into major or minor categories based on the length of time necessary to compile an answer. The major questions involve longer searches, bibliography etc. while the minor ones are usually factual and can be an hour or so. Overall the information enquiries dealt with by SILS are fewer than last year (314 as compared with 359 in total), but they still came from thirtyeight, mostly developing countries. It has also been seen that they originate from well defined organisations and groups of people, including ODM sponsored entomologists on Technical Cooperation projects abroad. By far the biggest group, as might be expected, are minor enquiries from COPR staff in UK (127). Academic and educational institutions in UK and overseas made similar numbers of enquiries, but with major enquiries in particular, those from commercial and non-commercial research organizations and from government departments abroad outnumbered those from similar bodies in UK.

The analysis of subjects raised, is that, understandably, the biology, ecology and control of pests, together with pesticides and spraying equipment form the bulk of the enquiries. An interesting fact, which also emerges is that 1 in 3 questions are about pests which are not currently the subject of research at COPR.

PHOTOGRAPHIC AND FILM COLLECTIONS

During the year 94 transactions, loans etc. were enacted involving 470 separate prints or transparencies. This was half as many again as the previous year and difficulty was found in keeping abreast of the demands, particularly when Mr D. Jay left in early August. Miss P. Wortley continued the servicing of enquiries, loans etc. The curation of the collection, however, was already far behind. Mr D. Jay and Miss P. Wortley managed to sort and annotate the individual collections and these are held in classified subject order as a temporary measure to enable them to be used. Fortunately, a number of excellent slides and photographs were deposited by the Centre staff in the collection.

The film collection was overhauled by Miss J. Reeve, vacation student, and all equipment was checked, together with existing loans. The projection room was used a little less than in 1974 (81 times as against 98).

DISTRIBUTION OF COPR PUBLICATIONS

The COPR publications list was updated to include price changes. SBNs and ISSNs were included for the first time and booksellers on our mailing list were circularised with copies. Mr W. F. Grant also arranged despatch of review copies of the *Hopper Development Atlas* and of gratis copies to several research organisations abroad.

Costs have been a source of concern during this year and postage particularly so. Fortyone reprints were distributed in two offers but other material was also offered, namely some duplicate FAO reports on locust research, some Anti-Locust reprints, some early issues of *Acridological Abstracts* and copies of the papers presented at the International Study Conference on the Current and Future Problems of Acridology in 1970. The total number of reprint requests was 934. Sixtynine sets of literature were sent to teachers and 398 requests were received from schools, colleges of education, polytechnics and universities.

STAFF AND ACCOMMODATION

The staffing situation continued throughout the year to be the major constraint severely restricting progress on the photograph collection and on the library indexing. Two long serving members of staff left, Mr D. Jay and Miss N. Gaze, as well as Miss M. Weld. Two members of staff from other divisions voluntarily worked overtime in SILS during November and December. Understandably, at the turn of the year some arears of work had accrued but the day to day services were kept up to date.

B PUBLICATIONS

The Centre's quarterly journal *PANS* was published regularly during 1975. In order to reduce its weight and economise on postage for its worldwide distribution a lighter weight paper was used in production from Volume 22 No. 2. Mr S. Mercer joined the *PANS* staff as technical editor in January.

PANS Manual No. 1, Pest Control in Bananas was translated into Spanish and published in 1975 as *Control de las Plagas de los Bananos*. 5000 copies were printed and 200 immediately distributed. About 500 copies of each of the other *PANS Manuals* was distributed during the year. Work on the revision of *PANS Manual No. 3, Pest Control in Rice* continued and it is expected that the new edition will be published in 1976.

In December, the twentyfirst year of publication of *PANS* was marked by a luncheon party attended by representatives of ODM, research organizations in UK and associated agro-industries.

The results of research undertaken by members of staff continued to appear mainly in the international scientific press; over 50 papers were published in 1975 (see Appendix 2). Accounts of work undertaken through ODM Research Schemes were published as special reports, two of them interim reports, on the Pesticides Residues Research Project with IITA in Nigeria and on the control of *Zonocerus variegatus* with the University of Ibadan in Nigeria, and a third on the problem of damage to sorghum by doves in Botswana. Seven Country Visit Reports gave details of advisory visits to various countries with recommendations to the governments concerned on further work. Two Miscellaneous Reports were also published.

The Report of the Centre for Overseas Pest Research for the period January to December 1974 was collated and edited by Dr B. Steele and gives an illustrated account of the Centre's activities throughout the year.

A *Tropical Pest Bulletin*, on the seasonal distribution and migrations of *Agrotis ipsilon*, a noctuid moth whose larvae can do considerable economic damage throughout the world, was also published. This bulletin resulted from a training programme undertaken at COPR by the author, P. Onyango Odiyo, on the biogeographical approach to pest forecasting, after which Mr Odiyo returned to EAAFRO and has been running the East African Armyworm Forecasting Service.

A new departure for the Publications Office was the preparation of a *Guide to Exhibits* for the Open Days in February. For this event also a new edition of the booklet *The Centre for Overseas Pest Research— Organisation and Projects* was produced, and later in the year yet another updated edition was published.

From time to time, the Publications Office is called upon to give advice and assistance to other organizations on matters concerning publication. One such case was the production of an issue of OICMA's non-periodic journal *Locusta*. COPR had already arranged the shortening and editing of a long paper by Dr R. A. Farrow on the ecology of the African Migratory Locust which resulted from his studies in Mali with OICMA in 1963—70. During 1975 the Publications Office, on behalf of OICMA, supervised the printing of this paper as *Locusta* no. 11 by an English printer, including marking up in the correct style and proof-reading this 200 page bulletin.

C INDUSTRIAL LIAISON

Friendly cooperation continued during this year between the Centre and many industrial companies both in UK and overseas. Dr H. S. Hopf, in his capacity as Industrial Liaison Officer paid and received a number of visits with representatives of these companies.

Information on industrial activities in the pesticide and application machinery fields is one of *PANS* requirements for its service to readers and commercial organizations regularly contribute in this way. In particular Miss S. C. A. Cook received assistance from a large number of commercial representatives in Central and South America during her tour to collect information for the revision of *PANS Manual No. 1, Pest Control in Bananas.*

COPR staff have also given advice to a number of commercial companies on pest control problems overseas. Subjects under enquiry included sugarcane and rice pests, soil pests in maize, banana pests and the use of commercial application machinery for granules and ultra-low-volume sprays in overseas situations.

Mr and Mrs D. R. Johnstone were involved in assisting firms with development of application machinery. In June they visited Micronair Ltd at Bembridge to discuss results of aerial spraying assessments they made in Swaziland and to obtain specifications and current performance data for the Micronair AU3000 atomiser. They discussed progress in the development of ultra-low volume spray machinery with Micron Sprayers at Bromyard in October and commenced a programme of laboratory testing of the new mini-ULVA sprayer currently under development.

Dr R. E. Roome gave advice to the Ciba-Geigy Agricultural Aviation Research Unit on *Heliothis* population and biology studies and supplied *Chilo partellus* pupae for testing sampling methods in UK for their work in the Sudan. Dr Roome also advised Tate and Lyle Ltd on control of cotton whitefly in Turkey.

Dr P. Matthiessen investigated for ICI the acute toxicity of a new candidate molluscicide against *Sarotherodon mossambicus.* It showed some promise in having a higher toxicity for snails than for fish.

Visits to COPR Termite Group for advice on preventing termite damage and materials testing were made by staff of several manufacturing companies. ICI Agricultural Division, Billingham, were interested in testing purlboard. Muirhead and Sons, timber engineers, of Grangemouth were concerned with the protection of shipping container floors. Tinsley Wire Industries, Sheffield, produce reinforced paperboard building panels. ICI of Jealotts Hill consulted the group in the course of the evaluation of pheromones in termite control.

Other requests for help and advice came from Dexion (Overseas) Ltd. (Termite-proofing benches for export to Middle East); Plessey Automation, Weybridge (PVC cables to be buried in Cuba); Hadrian Constructions, London Ltd. (House cladding for contract with Nigerian Government); Farmer and Dark, Architects (For houses in Magadiscio and Curazao); Plessey Navuids (Airfield lighting circuits for Zambia); Harding Associates Ltd., Leeds (Prefab housing for Middle East); Swiftplan Ltd., Southall (Prefab buildings for Persian Gulf); H. L. Dawson, Consulting Engineers (Protection of pipe insulation in Bandar Abbas); Watermeyer-Legge, Consulting Engineers (Polythene reservoir linings for Zambia); A. Sanderson & Sons Ltd., Christchurch, Hants (Luxury Wall coverings for Middle East); Secometric Ltd., Wickford, Essex (Prefab Hospital units for Nigeria); Plessey Radar (Protection of cables); Wm. Cowlin Joinery Ltd., Bristol (Prefab buildings for Bahrein); Ove Arup and Partners, London (Parliament Building dome in Riyadh, Saudi Arabia); Plessey Lighting (Fibre glass poles for circuits in Nigeria); G. E. C. Overseas Division, London (Protection of underground cables in Cuba); Shell International, London (Termite and rodent attack on plastics); British Wood Preserving Association, London (Protecting timber against termites, Iran); Halcrow Ltd. (Protection of timber crash barriers in Trucial States); Frazer-Nash Ltd., Hampton Wick (Cables in tropical soils); Dow Chemicals (U.K.) Ltd., London (Feasibility of House fumigation in Africa against *Cryptotermes*); Decca Navigation Co. (Possible damage to copper coating of aerials).

Mr R. M. C. Williams, also of the Termite Group has completed laboratory trials of the termiticidal properties of the substances PH60-38 and PH60-40 produced by Philips-Duphar B.V. of Holland and screening tests against termite attack on polythene foam for Hewitt and Cox Ltd. of Gloucester. Screening tests of cable sheaths for Pirelli General Ltd., Eastleigh, Hants, have continued during the year.

Mr E. G. Harris has received valuable cooperation on evaluation of acaricide resistance test methods from Wellcome Research Laboratories, who have supplied batches of ticks regularly for bioassays.

There has been considerable cooperation from industry in the development of the armyworm pheromone testing programme. ICI Plant Protection have provided experimental samples of microencapsulated pheromone for evaluation in Crete and Mr N. A. Hunt and Mr R. Henry of Food Industries Ltd. paid visits to the island during experimental application to further cooperation in the commercial production of these materials.

D ECONOMIC SALTATORIA

Mrs J. MacDonald continued her work on the Index of Economic Saltatoria and, by the end of 1975 over 2350 published works and a large number of unpublished reports had been scanned for information on the species causing, or likely to cause, crop damage. In March 1975 work commenced on summarizing the information for each species and by the end of the year just over 100 species out of a total of about 560 had been written up for publication.

E INFORMATION LECTURES

Lectures by COPR staff for their colleagues were not organized regularly during 1975. The following list is therefore somewhat smaller than in the previous year:

12 February	Migrations by successive generations of African armyworms in Eastern Africa and the south-western Arabian peninsula. Miss E. Betts.
12 June	Meteorology and *Simulium* migration. Mr D. Pedgley.
18 June	Selection of oviposition sites by adult stalkborers. Dr R. E. Roome.
21 October	Observations of the structure of sense organs on the ovipositor tip of *Chilo partellus* by the scanning and transmission electron microscope. Mrs G. Chadha.
2 December	Research-based recommendations versus farmer practice — some lessons from cotton spraying in Malawi. Mr J. Farrington.

In addition, several of the Centre's research groups organised day symposia in which their research programmes were explained and subjected to scrutiny by fellow staff and external advisers.

VI TRAINING

A COURSES AND LECTURES

The activities of COPR staff have always included participation in and the promotion of training of UK and overseas personnel in the numerous aspects of applied entomology and crop protection.

Overseas projects invariably involve local training and the placement of counterparts who require specialised training at COPR or UK universities which offer special Crop Protection Courses at graduate and post-graduate levels.

In addition requests for training of extension workers, agricultural officers etc have resulted in COPR staff carrying out short term visits to overseas countries to demonstrate and lecture on the most modern techniques available. Organizations such as FAO, WHO have also requested the services of COPR staff for a variety of courses sponsored in developing countries, and several UK universities receive specifically designed short term courses for graduates and post-graduates in which typical examples of the problems of crop protection in overseas countries are demonstrated.

There were encouraging signs during 1975 of an increased demand for the services of COPR staff in organising and supporting training courses and extension exercises overseas. Mr G. B. Popov assisted in two FAO/UNDP courses on crop pest control with special reference to Desert Locust control and research. He was technical co-ordinator/organiser for a course in Dakar (Senegal) from 17 February to 21 March. He lectured on (1) General introduction to the locust problem; (2) Airborne insects and atmospheric environment (on behalf of Dr Rainey); (3) The biogeographical basis of warning services and control organisations (on behalf of Miss Betts); (4) Desert Locust survey techniques and estimation of populations and (5) Trapping techniques as employed for population studies and control.

The second course was in Algiers from 8 to 24 November. Mr Popov lectured on (1) Biology and phases in locusts; (2) Locusts in plagues and in recessions; (3) Methods of surveying and assessing recession populations and (4) The economic importance of grasshoppers in the Sahelian zone. Both these courses were conducted in French.

Organized by FAO, a very successful workshop on crop pest and loss assessment was held at the University of Bari, Italy, in April. Mr P. T. Walker, who helped to plan the course, gave lectures on the assessment of pests, on methods of carrying out trials on pest-loss correlation, and on the various relationships found between crop losses and pest attack. Other lectures were given on losses caused by diseases in plants, and the part played by crop loss assessment in decision making in pest management. Field excursions were made to examine pests in the field, including cereal, olive and beet pests. The workshop was well attended by workers from 15 countries, and individual problems were discussed. It is expected that several more workshops will be requested by overseas countries as a result.

At the invitation of the Government of Kenya, and following some years research on maize pests in Kenya, Mr P. T. Walker, Mr B. W. Bettany and Mr M. J. Hodson, visited Kenya in April and May on an extension tour of the main maize growing areas. Mr Bettany and Mr Hodson had made a sound film, colour, on the maize stem borer, *Busseola fusca*, in Kenya, including details of its life cycle, and how to control it by chemical and other means. A version of the film in Swahili has just been released. The film was shown, sometimes out of doors at night, sometimes in community halls or farmers' centres, and talks, sometimes in Swahili, sometimes translated into other local languages, were given. Talks were illustrated by pictures, specimens and attacked plants, and demonstrations with insecticide granules were given in the field. In all, 48 sessions were held, reaching an estimated 2000 people, and 1000 leaflets were distributed. More information on better maize growing was asked for, but it was felt that the film and lecture would make more impact if limited to its specialized subject. A report was issued analyzing the logistics of the exercise, the difficulties and possible improvements which could be made in a future tour.

A course of 20 lectures, concluding in April, was given on tropical soils by Dr T. G. Wood, whilst representing ODM at the Agricultural Research Station, Mokwa, Nigeria, for prospective candidates in the City and Guilds examination in Tropical Agriculture.

A two-day course on mapping and surveying was held in early June by Mr and Mrs R. A. Steedman at the Locust Control Research Centre at Chaibadan, Thailand. It was attended by heads of the field survey teams and by members of the Agricultural Aviation Division.

Dr J. I. Magor and Mr D. E. Pedgley arranged a seminar for entomologists of the WHO Onchocerciasis Control Programme and scientists from ORSTOM and ASECNA at Bobo Dioulasso, Upper Volta on 5 May on the subject of "Insect movement and weather".

Dr W. Sands gave a seminar to staff and students of Kenyatta College, Nairobi in June on the distribution of East African termites. He also gave two seminars in Sri Lanka to staff at the Tea Research Institute, Talawakele and staff and students at Colombo University on "Termites as tree and crop pests", both in October.

In Botswana, Mr G. G. Pope gave a lecture to staff of the Botswana Weather Bureau entitled "Birds and the weather" as part of a seminar organised to give staff a wider appreciation of the overall effects of the weather.

Four days in March were devoted to the annual short course for M.Sc. students at Imperial College, arranged by Dr P. Ellis. She also lectured on "The theory of locust phases", "Prospects for the biological and ecological control of locusts" and "*Spodoptera littoralis* — can we control this pest with pheromones?" Other members of COPR who lectured included Mr T. Thompson on practical and standard methods of testing locusts with pesticides and a lecture on the Quelea bird problem; Miss L. Bennett "Biogeographical analysis in the study of pests"; Mr W. Baker "Flight behaviour of locusts"; Mr M. Lambert "Flight in the Australian locust"; Mr C. Lee "Control of mosquitoes in the Cayman Islands"; Mr J. Perfect "Pesticides and soil ecology in the tropics"; Dr W. Sands and Dr R. M. Williams gave lectures and demonstrations on termites, Dr P. Mathiessen on "Testing molluscicides on fish" and Dr J. Duncan, "The Schistosomiasis problem"; Miss E. Betts, "The armyworm problem in eastern Africa"; Dr R. Roome, "Plant resistance with special reference to lepidopterous pests"; Mr D. McKinley, "Viruses in insect control" and Miss A. Lumley demonstrated the COPR library and information service.

On 10 and 11 June COPR organised two days of lectures during the six week ODM/ Imperial College course on Pest Management, attended by ten overseas students.

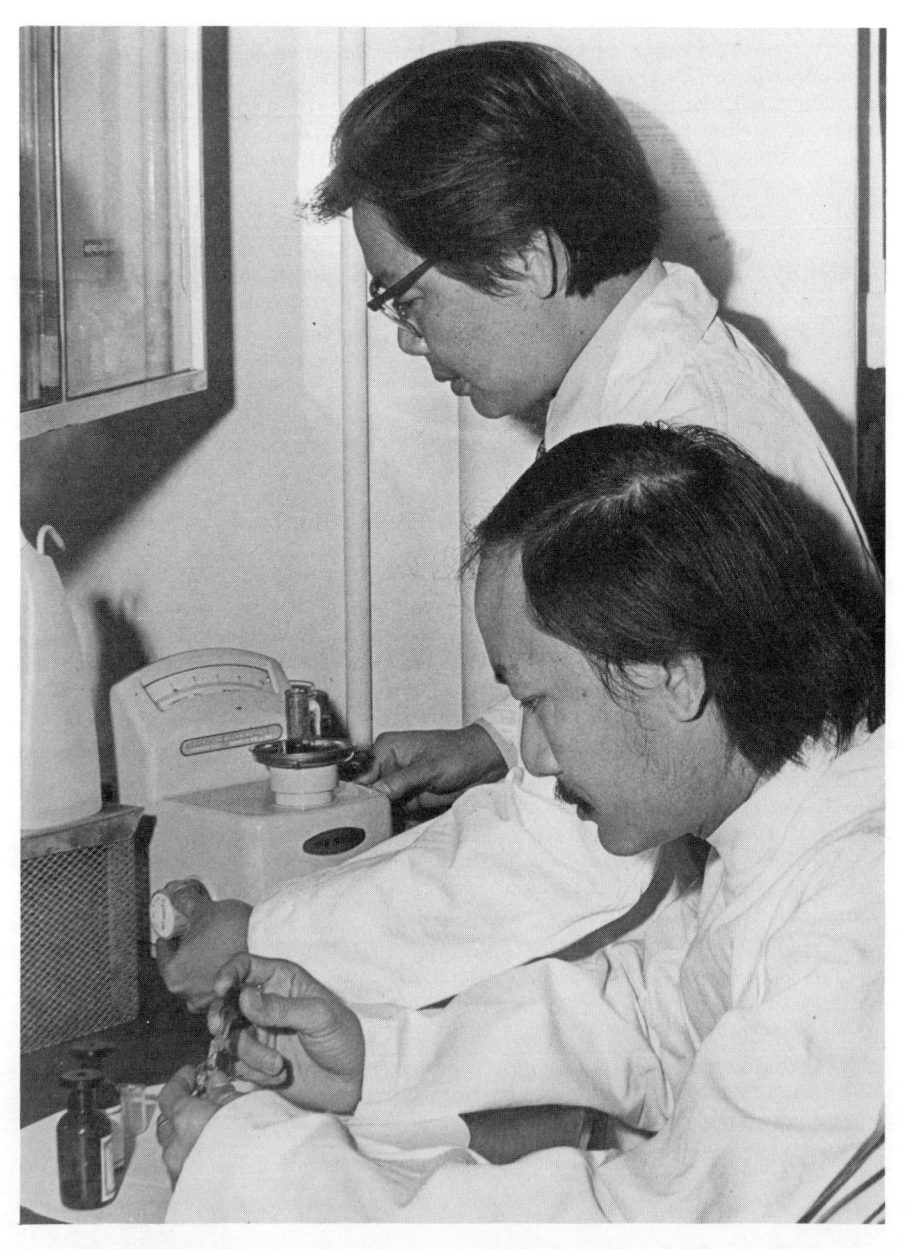

Fig. 28. Mr. S. Ratanavong from Laos (above) and Mr. Songwood Pojananuwong from Thailand, both of whom have carried out research projects at COPR as part of their training in UK.

These lectures also were arranged by Dr P. Ellis who spoke on "Use of pheromones in insect control" and "Prospects for the ecological and biological control of locusts". Amongst other members of staff, Dr R. Chapman lectured on "Is *Zonocerus* a pest in Nigeria?"; Dr R. Rainey on "Control of locusts and migrant pests with particular reference to meteorology"; Dr P. Hunter-Jones on "Special features of chemical control in locusts"; Dr H. S. Hopf on "Some rodent problems"; Miss E. Betts on "The armyworm problem in Eastern Africa"; Mr T. Thompson on "The Quelea bird problem" and Mr R. M. C. Williams on "The termite problem". Miss P. Wortley and Miss A. Lumley arranged a tour of the COPR library and information service.

In February Dr R. Chapman assisted by Mr T. Perfect organised an informal 3-day workshop at COPR on "Implications of pesticide use for tropical freshwater and terrestrial ecosystems" and this was published as the first of the new COPR Proceedings series.

Special training courses in pest control were again organised for four universities, Strathclyde, Bangor, Bath and Salford and several members of COPR staff took part. Dr R. F. Chapman is also Visiting Professor at the University of Hull where he gave a course of lectures on locust biology and two lectures on the biology and control of *Zonocerus variegatus* as part of a course in applied biology. In November Dr P. Ellis conducted an afternoon symposium at Hull University on insect sex pheromones.

In the field of meteorology, Mr D. Pedgley gave a lecture on "Weather of Ethiopia" for the Meteorological Office course on tropical meteorology, arranged a week's field course at Drapers' Field Centre, N. Wales, sponsored by the Royal Meteorological Society on "Mountain weather" and lectured to the London University College, Geographical Department seminar on "Why does it rain in the desert". He also organised a course for the Reading geophysics seminar in which he was assisted by Mrs A. Steedman.

Dr R. E. Roome gave a lecture to the University of East Anglia School of Developmental Studies on "Crop resistance with reference to subsistence crops".
Dr M. R. K. Lambert addressed the Zoological Society of University College, London on "The flight activity of locusts" and Dr A. B. Hadaway gave a lecture on "Insecticide formulations" at the courses on "Aerial application of insecticides" at the Cranfield Institute of Technology.

Miss A. Lumley organised an afternoon training programme for a party of Library Students from Leeds Polytechnic in October. The programme carried out with the assistance of Miss P. Wortley demonstrated the work of SILS as an example of a "special government library".

Mr A. Antoniou maintained during the year the service of advice to Universities, Colleges and Schools on rearing and breeding locusts. This was of assistance in establishing and maintaining health cultures of locusts which are now widely used for research and teaching purposes.

B VISITING RESEARCH WORKERS

As in previous years, a large number of visitors from overseas and students working on problems of pest control appropriate to developing countries spent varying periods of time studying at the Centre and benefitting from the experience of the Centre's staff.

In November, Mr W. Namurangsri of the Ministry of Agriculture and Cooperatives in Thailand began studies at the Centre under Mrs A. Steedman. He is being sponsored by ODM to work on the distribution of *Patanga succincta*, a serious pest in his own country.

Dr A. Navon from the Volcani Institute in Israel began a year as a visiting research worker. He studied the non-protein amino acids and their effect on feeding and development of locusts.

Mr Asim A/R Dafalla continued his three years secondment to work with the COPR Molluscicide Group with the support of the Sudanese National Research Council. He is registered with the London School of Hygiene and Tropical Medicine for a London Ph.D.

Mr C. E. Ohiagu of Ahmadu Bello University in Nigeria worked during 1974 and 1975 on grass-harvesting termites in cooperation with COPR staff in Nigeria. At the end of the year he started to write his thesis for submission for his Ph.D. degree in 1976.

Dr D. Campion continued his supervision of Mr Andreas Krambias of the Ministry of Agriculture in Cyprus for his Ph.D. degree, as did Dr Jago for Mr J. E. Ohabuicke of OICMA. The latter visited the Centre for a few months to finalise his thesis on nutrition and egg resorption in the African Migratory Locust, which was successfully submitted to the University of Reading and he was awarded his Ph.D. degree in December.

Miss A. Lumley supervised a specially arranged course on information work for Mr A. S. K. Atsa of Ghana prior to his undertaking an M.Sc. course in Information Science at the City University in London.

As usual, a number of visitors were assisted with short courses by Dr W. A. Sands of the Termite Unit. Mr Tho Yow Pong of the Malaysian Forest Research Institute spent twelve days in September studying the taxonomy of Malaysian termites. Mr Hassan Osman El Nour of the Sudan received a short course in termite taxonomy for two weeks in May. Dr G. Rohrmann of the University of Botswana, Lesotho and Swaziland did a literature survey in connection with his work on termite energetics and Mr J. Adetunji from Ilorin, Nigeria a survey for a dissertation on timber termites. Dr D. Pomeroy of Kenyatta College, Nairobi studied the taxonomy of East African *Odontotermes* during a short visit.

Two vacation students are noted as having worked with the Centre during the year; Mr B. Seymour-Taylor of Bristol Polytechnic worked on insect breeding with the Division of Chemical Control at Porton Down and Mr D. Morgan of the University of Leicester was with the Division of Biology in London investigating the relative importance of various factors in the development of the characteristic green colour of isolated Desert Locust hoppers.

A number of Sandwich Course students spent six months to one year at COPR. Mr Ian Middelfel of Brunel University assisted the work of the Division of Biology on water relations and drinking behaviour in locusts and also carried out bioassays of insecticides against locusts. Mr R. Dawson of Sunderland Polytechnic worked initially on the culturing of locusts and subsequently on crossing, self-crossing and back-crossing two races of *Locusta migratoria*. Mr B. Long of Brunel University worked for six months on turning responses during locust flight.

Mr S. V. Brown of Sunderland Polytechnic completed his year's training with the Molluscicide Unit and was replaced by Mr C. M. Fox from Trent Polytechnic. Two students worked with Dr P. E. Ellis during the year. Mr J. Butcher of Bradford Polytechnic received training in the management of locust and caterpillar stocks. He started a special project on the water gain and loss of the caterpillars in relation to their moulting cycle. Mr R. Ash studied the laying preferences of *Spodoptera littoralis* females. Preferences between certain leaves were shown to be dependent on sense organs of the antennae. Electron microscope and electro-physiological studies of the ovipositor showed that there were both mechanical and chemosensory hairs present.

APPENDICES

1 STAFF OF THE CENTRE FOR OVERSEAS PEST RESEARCH*

Director: P. T. Haskell, CMG, Ph.D.   Chief Scientific Officer (B)
Personal Secretary: Miss M. Murphy
Deputy Director: T. Jones, OBE, B.Sc.   Deputy Chief Scientific Officer (1)
Personal Secretary: Vacant

OFFICE OF THE DIRECTOR

H. S. Hopf, Ph.D.	Scientific Secretary/Principal Scientific Officer
C. F. Hemming, M.A.	Scientific Secretary/Principal Scientific Officer
C. W. Lee, B.Sc.	Scientific Secretary/Principal Scientific Officer

Administration Group

C. F. G. Foss	Administrative Secretary/Senior Executive Officer
Miss D. L. Moors	Higher Executive Officer
I. P. Grant	Higher Executive Officer (Supernumerary)
F. A. Lee	Higher Scientific Officer
J. Morgan-Smith	Higher Scientific Officer
H. Woodman	Executive Officer
A. I. Cossins	Executive Officer (Supernumerary)
Miss S. Koller	Clerical Officer
N. G. Hurst	Clerical Officer
Miss B. Lee	Clerical Officer
A. Munday	Clerical Officer
Mrs M. Slater	Paperkeeper
Mrs M. F. Payne	Messenger
P. Miller	Messenger
Mrs J. Davidson	Messenger
	3 typists

Scientific Information and Library Service

Miss P. J. Wortley, B.Sc.	Senior Scientific Officer
B. Steele, Ph.D.	Senior Scientific Officer
Miss J. E. R. Salter, M.A.	Senior Scientific Officer
Miss M. A. Ward, B.Sc.	Senior Scientific Officer
Miss J. M. Child, B.Sc.	Senior Scientific Officer
Mrs J. MacDonald, B.Sc.	Higher Scientific Officer
Miss A. Lumley, B.Sc.	Higher Scientific Officer
K. Smith	Higher Scientific Officer
Mrs J. S. Ridout, B.Sc.	Higher Scientific Officer
Vacant	Higher Scientific Officer
Miss E. J. Luard, M.Sc.	Scientific Officer
Miss S. C. A. Cook, B.Sc.	Scientific Officer
S. L. Mercer, B.Sc.	Scientific Officer
Miss P. Schofield, B.A., M.Sc.	Scientific Officer
Miss F. K. Sherwood, B.Sc.	Scientific Officer

---

*This list refers to COPR Staff as at 31 December 1975. Staff changes occurred during 1975 so some staff mentioned elsewhere in this Report may not appear in the list.

Vacant	Scientific Officer
Mrs A. Walpole	Higher Grade Cartographer (part time)
Miss H. Loewy	Clerical Officer
P. Shannon	Clerical Officer
W. F. Grant	Clerical Officer
Miss J. E. Butler	Clerical Officer (part time)

DIVISION OF CHEMICAL CONTROL

A. B. Hadaway, Ph.D.	Assistant Director/Senior Principal Scientific Officer
F. Barlow, B.Sc.	Principal Scientific Officer
D. R. Johnstone, B.Sc.	Principal Scientific Officer
Vacant	Principal Scientific Officer
M. P. L. Fogden, D.Phil.	Senior Scientific Officer[1]
W. S. Watts	Senior Scientific Officer
D. J. McKinley, M.Sc.	Senior Scientific Officer
A. T. Thompson, B.Sc.	Senior Scientific Officer
N. S. Irving, B.Sc.	Senior Scientific Officer (Home Based)[2]
E. G. Harris, B.Tech.	Senior Scientific Officer
Vacant	Senior Scientific Officer
G. G. Pope	Higher Scientific Officer
C. R. Turner	Higher Scientific Officer
J. E. H. Grose	Higher Scientific Officer
L. S. Flower	Higher Scientific Officer
Mrs K. A. Johnstone	Higher Scientific Officer
W. J. King	Higher Scientific Officer
Mrs V. Forder	Assistant Scientific Officer
Miss S. C. Smith	Assistant Scientific Officer
Vacant	Assistant Scientific Officer
Mrs V. M. D. Haines	Typist
Vacant	Research and Development Craftsman
Mrs S. Hill	Experimental Worker
R. P. Smith	Experimental Worker

DIVISION OF ECOLOGY

C. Ashall, B.Sc.	Assistant Director/Senior Principal Scientific Officer
R. C. Rainey, D.Sc.	Senior Principal Scientific Officer
G. B. Popov, MBE	Principal Scientific Officer
D. E. Pedgley, B.Sc.	Principal Scientific Officer
J. R. Riley, Ph.D.	Principal Scientific Officer
P. T. Walker, M.Sc.	Principal Scientific Officer
J. Roffey, B.Sc.	Principal Scientific Officer
Miss S. M. Green, M.A.	Principal Scientific Officer
Miss J. I. Magor, Ph.D.	Principal Scientific Officer
Miss E. Betts, B.A.	Senior Scientific Officer
Miss L. J. Rosenberg, Ph.D.	Senior Scientific Officer
R. J. Douthwaite, M.Sc.	Senior Scientific Officer
J. Murlis, Ph.D.	Senior Scientific Officer
J. Tunstall	Senior Scientific Officer
J. Farrington, M.Phil.	Senior Scientific Officer

---

[1] On secondment to Government of Mexico
[2] On overseas service in West Indies

Vacant	Senior Scientific Officer
I. B. Jones	Higher Scientific Officer
Miss M. J. Haggis	Higher Scientific Officer
R. A. Steedman, B.Sc.	Higher Scientific Officer
Mrs A. Steedman, B.Sc.	Higher Scientific Officer
J. R. W. Harris, D.Phil.	Higher Scientific Officer
D. R. Reynolds, Ph.D.	Higher Scientific Officer
M. R. K. Lambert, Ph.D.	Higher Scientific Officer
Vacant	Higher Scientific Officer
M. R. Tucker, B.Sc.	Scientific Officer
Miss E. M. Paton	Scientific Officer
Vacant	Scientific Officer
B. W. Bettany	Assistant Scientific Officer
Miss J. de Leon	Assistant Scientific Officer

DIVISION OF BIOLOGY

R. F. Chapman, D.Sc.	Assistant Director/Senior Principal Scientific Officer
W. A. Sands, D.Sc.	Senior Principal Scientific Officer
Miss P. E. Ellis, D.Sc.	Principal Scientific Officer
R. M. Williams, M.Sc.	Principal Scientific Officer
J. C. Davies, Ph.D.	Principal Scientific Officer (Home Based)[3]
W. R. Ingram, B.Sc.	Principal Scientific Officer (Home Based)[4]
A. B. S. King, Ph.D.	Principal Scientific Officer (Home Based)[5]
D. G. Campion, Ph.D.	Principal Scientific Officer
P. Ward, Ph.D.	Principal Scientific Officer (Home Based)
T. G. Wood, Ph.D.	Principal Scientific Officer
N. D. Jago, Ph.D.	Principal Scientific Officer
J. Duncan, Ph.D.	Principal Scientific Officer (Home Based)
P. Hunter-Jones, Ph.D.	Principal Scientific Officer
J. E. Moorhouse, Ph.D.	Principal Scientific Officer[6]
R. E. Roome, Ph.D.	Senior Scientific Officer (Home Based)
P. H. Goll, M.Sc.	Senior Scientific Officer[7]
Miss E. A. Bernays, Ph.D.	Senior Scientific Officer
P. J. Jones, B.Sc.	Senior Scientific Officer (Home Based)[8]
G. A. Mitchell, Ph.D.	Senior Scientific Officer[9]
R. J. Cooter, Ph.D.	Senior Scientific Officer
T. J. Perfect, M.Sc.	Senior Scientific Officer
A. Antoniou	Senior Scientific Officer
A. R. McCaffery, Ph.D.	Senior Scientific Officer
G. G. Cavanagh, B.Sc.	Higher Scientific Officer
B. R. Critchley, Ph.D	Higher Scientific Officer[10]
R. Yeadon, B.Sc.	Higher Scientific Officer[10]
P. Mathiessen, Ph.D.	Higher Scientific Officer
J. W. M. Logan, B.Sc.	Higher Scientific Officer
Miss J. B. Mason	Higher Scientific Officer

---

[3] On secondment to ICRISAT, Hyderabad.
[4] On overseas service in the Eastern Caribbean.
[5] On secondment to the Government of Costa Rica.
[6] On secondment to ARC Imperial College Field Station, Silwood Park.
[7] On secondment to Bilharzia project, The Gambia.
[8] On overseas service with the Bird Pest Project, Nigeria.
[9] On overseas service in the Eastern Caribbean.

D. E. Padgham, M.Sc.	Scientific Officer
W. W. Page	Scientific Officer[10]
Miss A. G. Cook, B.Sc.	Scientific Officer
S. Bacchus, B.Sc.	Scientific Officer
A. L. Davies	Scientific Officer
L. J. McVeigh	Scientific Officer
J. H. Davies	Scientific Officer
Miss E. M. Leather	Assistant Scientific Officer
N. A. Hannaway	Assistant Scientific Officer
D. J. Chamberlain	Assistant Scientific Officer
Miss I. P. B. McAleer	Assistant Scientific Officer
Mrs G. K. Chadha	Assistant Scientific Officer
R. W. Lamb	Assistant Scientific Officer
M. J. Pearce	Assistant Scientific Officer
Miss N. Brown	Assistant Scientific Officer
Mrs. S. E. J. Storer	Assistant Scientific Officer
Mrs G. A. Colquhoun	Photographer
A. Few	Laboratory Attendant
Mrs. P. M. Corrigan	Laboratory Attendant
Mrs J. Blaylock	Laboratory Attendant
Miss S. Coggin	Laboratory Attendant
Miss A. Rabe	Laboratory Attendant
E. G. Shurmer	Laboratory Mechanic
Vacant	Experimental Worker Grade IV

STAFF ATTACHED TO COPR

Fellowships

D. W. Ewer, Ph.D.	Principal Research Fellow
Mrs E. M. Home, Ph.D.	Junior Research Fellow
Miss M. M. Jones, B.Sc.	Junior Research Fellow
R. W. Dunlop, B.Sc.	Junior Research Fellow
J. M. Ritchie, B.Sc.	Junior Research Fellow
N. M. Collins	Junior Research Fellow
P. S. Baker, B.Sc.	Junior Research Fellow
J. Draper, B.Sc.	Junior Research Fellow

Sandwich Course Students

P. Scott
C. M. Fox
J. Butcher
P. Gregory

Other associated staff

A. Russell-Smith, M.Sc.	ODM Environmental Scheme R2730
A. D. Smith	ODM Radar Research Project R2970
J. S. S. Beesley	ODM Bird Pest Project R2664
Mrs U. Critchley	ODM Environmental Scheme R2730
R. A. Johnson, Ph.D.	ODM Termite Project R2709
Asim A/R Dafalla	Seconded from LSH&TM
P. Rosen, Ph.D.	ODM Pest and Vector Control Project R3069

---

[10] On overseas service with the Environmental Scheme, IITA, Nigeria.

2 PUBLICATIONS OF THE CENTRE FOR OVERSEAS PEST RESEARCH

PERIODICAL PUBLICATIONS

Acridological Abstracts

1974: Nos 1–2 (Abstracts 9201–9400); Nos 3–4 (Abstracts 9401–9600); Subject Index to Abstracts 8001–9000.

1975: Nos 1–2 (Abstracts 9601–9800).

Acridological Abstracts (New Series) (Published in *Acrida*)

1974: Part 1 (No. 1–201); Part 2 (No. 202–381); Part 3 (No. 382–560); Part 4 (No. 561–745).

1975: Part 5 (No. 746–862); Part 6 (No. 863–1062); Part 7 (No. 1063–1221); Part 8 (No. 1222–1408).

PANS (Pest Articles and News Summaries)

A quarterly journal on all aspects of tropical pest control including plant diseases and weeds. Subscription rate £12 per annum post free by surface mail, airmail subscription £19 per annum. Issues published during 1974: Volume 21 parts 1–4.

PANS Cumulative Index, Vols 1–18, 1955–1972. Price £1.20.

PANS Manuals

No. 1. Control de las plagas de los bananas. (Spanish edition of 'Pest control in Bananas'.) Price 80p.

Tropical Pest Bulletins

No. 4. Seasonal distribution and migrations of *Agrotis ipsilon* (Hufnagel) (Lepidoptera, Noctuidae). P.O. Odiyo, 1975. 26pp. Price 75p.

SCIENTIFIC PAPERS

BACCHUS, S. and KENDALL, M. D. (1975). Histological changes associated with enlargement and regression of the thymus glands of the red-billed quelea *Quelea quelea* L. (Ploceidae: weaver-birds). *Phil. Trans. R. Soc. Lond.*, (B), 273: 65–78.

BAKER, P. S. (1975). Optomotor responses of flying locusts.* [Abstr.] *Exp. Brain Res.*, 23 (Suppl.): 13. [Abstr. No. 22.]
 *Paper presented at the First European Neurosciences Meeting, Munich, September 9–13, 1975.

BARLOW, F. and HADAWAY, A. B. (1975). The insecticidal activity of some synthetic pyrethroids against mosquitoes and flies.* *PANS,* 21: 233–238.
 *Paper presented at the Third International Congress of Pesticide Chemistry, Helsinki, 1974.

BARTON BROWNE, L., MOORHOUSE, J. E. and VAN GERWEN, A. C. M. (1975). Sensory adaptation and the regulation of meal size in the Australian plague locust, *Chortoicetes terminifera*. *J. Insect Physiol.*, 21: 1633–1639.

BARTON BROWNE, L., MOORHOUSE, J. E. and VAN GERWEN, A. C. M. (1975). An excitatory state generated during feeding in the locust, *Chortoicetes terminifera*. *J. Insect Physiol.*, 21: 1731–1735.

BENNETT, L. V. (1975). Development of a desert locust plague. *Nature, Lond.*, 256: 486–487.

BERNAYS, E. A., BLANEY, W. M. and CHAPMAN, R. F. (1975). The problems of perception of leaf-surface chemicals by locust contact chemoreceptors. pp. 227—229. *In*: Denton, D. A. & Coglan, J. P. (eds.) *Olfaction and taste. Volume 5.* London, Academic Press.

BERNAYS, E. A. and CHAPMAN, R. F. (1975). The importance of chemical inhibition of feeding in host-plant selection by *Chorthippus parallelus* (Zetterstedt). *Acrida*, 4: 83—93. [With French summary.]

BERNAYS, E. A., CHAPMAN, R. F., COOK, A. G., MCVEIGH, L. J. and PAGE, W. W. (1975). Food plants in the survival and development of *Zonocerus variegatus* (L.). *Acrida*, 4: 33—45. [With French summary.]

CAMPION, D. G. (1975). Chemosterilants for *Diparopsis castanea.* PANS*, 21: 359—364.
*Paper read at the International Congress of Pesticide Chemistry, 3rd, Helsinki, 1974.

CAMPION, D. G. (1975). Insect chemosterilants: the present situation. *Bull. Rech. agron. Gembloux*, 5 (Special): 171—179.

CAMPION, D. G. (1975). The potential use of pheromones in insect control. *Proc. Br. Insectic. Fungic. Conf.*, 8th(3): 939—946.

CAMPION, D. G. (1975). Sex pheromones and their uses for control of insects of the genus *Spodoptera. Meded. Fac. LandbWet. RijksUniv. Gent*, 40: 283—292.

CAMPION, D. G. (1975). The sex pheromone of the cotton leafworm *Spodoptera littoralis* Boisd. *Int. Cong. Pl. Prot.*, 8th, Moscow, Rep. Inf. Section V: 48—49.

CAMPION, D. G., BETTANY, B. W., NESBITT, B. F., BEEVOR, P. S., LESTER, R. and POPPI, R. G. (1975). The synthetic sex-pheromone of *Spodoptera littoralis* Boisd. and its uses. A field evaluation. pp. 593—609. *In*: International Atomic Energy Agency & FAO, *Sterility principle for insect control 1974. Proceedings of the symposium on the sterility principle for insect control jointly organized by the International Atomic Energy Agency and the Food and Agriculture Organization of the United Nations, and held in Innsbruck, 22—26 July, 1974*. Vienna, International Atomic Energy Agency. (Proceedings Series.)

CHAPMAN, R. F. (1975). *Zonocerus variegatus* (L.), an enigmatic grasshopper. [Abstr.] *Proc. R. ent. Soc. Lond.*, 39: 21. [Discussion: *ibid.*, 39: 27—28.]

COOTER, R. J. (1975). Ocellus and ocellar nerves of *Periplaneta americana* L. (Orthoptera: Dictyoptera) *Int. J. Insect Morph. Embryol.*, 4: 273—288.

COOTER, R. J. (1975). Visually stimulated behaviour and locust flight.* [Abstr.] *Exp. Brain Res.*, 23(Suppl.): 43. [Abstr. no. 82].
*Paper presented at the first European Neurosciences Meeting, Munich, September 9—13, 1975.

DAVIES, J. C. (1975). Insecticides for the control of the spread of groundnut rosette disease in Uganda. *PANS*, 21: 1—8.

DAVIES, J. C. (1975). Use of menazon insecticide for control of rosette disease of groundnuts in Uganda. *Trop. Agric.*, 52: 359—367.

FARROW, R. A. (1975). The African migratory locust in its main outbreak area of the Middle Niger: quantitative studies of solitary populations in relation to environmental factors. *Locusta*, no. 11: 198pp.

GOODMAN, L. J., PATTERSON, J. A. and MOBBS, P. G. (1975). The projection of ocellar neurons within the brain of the locust, *Schistocerca gregaria. Cell Tissue Res.*, 157: 467—492.

HADAWAY, A. B. (1975). Keynote address. pp.3—8. *In: The CENTO seminar on the toxicology of pesticides with special reference to environmental hazards, held in Tehran, October 1974.* Ankara, Turkey, Central Treaty Organisation.

HASKELL, P. T. (1975). Britain's role in international pest control. *Proc. Br. Insectic. Fungi. Conf.*, 8th (3): 1059—1067.

HOME, E. M. (1975). Ultrastructural studies of development and light-dark adaptation of the eye of *Coccinella septempunctata* L., with particular reference to ciliary structures. *Tiss. Cell*, 7: 703—722.

INGRAM, W. R. (1975). Improving control of the vegetable armyworm. *PANS,* 21: 162–167.

JOHNSTONE, D. R., HUNTINGTON, K. A. and KING, W. J. (1975). Development of hand spray equipment for applying fungicides to control *Cercospora* disease of groundnuts in Malawi. *J. agric. Engng Res.,* 20: 379–389.

JOHNSTONE, D. R., HUNTINGTON, K. A. and QUINN, J. G. (1975). Physical assessment of very low volume fungicide spray application on tomatoes in northern Nigeria. *Proc. Br. Insectic. Fungic. Conf.,* 8th (1): 175–182.

JONES, P. J. (1975). The significance of bird migration to bird control strategy. *Proc. Br. Insectic. Fungic. Conf.,* 8th (3): 883–889.

KENDALL, M. D. and WARD, P. (1975). Thymus glands and erythropoiesis. p. 372 *In*: Yamada, E. (ed.), *Proceedings of the 10th. International Congress of Anatomists and the 80th Annual Meeting of the Japanese Association of Anatomists, Tokyo, 1975.* Tokyo, Science Council of Japan.

KING, A. B. S. (1975). The extraction, distribution and sampling of eggs of the sugar-cane froghopper, *Aeneolamia varia saccharina* (Dist.) (Homoptera, Cercopidae). *Bull. ent. Res.,* 65: 157–164.

KING, A. B. S. (1975). Factors affecting the phenology of the first brood of the sugar-cane froghopper *Aenolamia varia saccharina* (Dist.) (Homoptera, Cercopidae) in Trinidad. *Bull. ent. Res.,* 65: 359–372.

LEE, C. W. (1975). Pyrethrum airsprays control mosquitoes in the Caribbean. *Pyrethrum Post,* 13: 5–6.

LEE, C. W., PARKER, J. D. PHILIPPON, B. and BALDRY, D. A. T. (1975). A prototype rapid release system for the aerial application of larvicide to control *Simulium damnosum* Theo. *PANS,* 21: 92–102.

MCCAFFERY, A. R. (1975). Food quality and quantity in relation to egg production in *Locusta migratoria migratorioides. J. Insect Physiol.,* 21: 1531–1534.

MCCAFFERY, A. R. and HIGHNAM, K. C. (1975). Effects of corpora allata on the activity of the cerebral neurosecretory system of *Locusta migratoria migratorioides* R. & F. *Gen. Comp. Endocr.,* 25: 358–372.

MCCAFFERY, A. R. and HIGHNAM, K. C. (1975). Effects of corpus allatum hormone and its mimics on the activity of the cerebral neurosecretory system of *Locusta migratoria migratorioides* R. & F. *Gen. Comp. Endocr.,* 25: 373–386.

MACCUAIG, R. D. (1975). [Unpubl.]. Cholinesterase levels in staff of Plant Protection Departments in Eastern Africa. *Joint FAO/IAEA Inf. Circ.,* no. 19: Abstr. no. 19.

MCCUAIG, R. D. (1975) [Unpubl.]. Depression of cholinesterase activity of domestic animals fed on insecticide treated vegetation. *Joint FAO/IAEA Inf. Circ.,* no. 19: Abstr. no. 20.

MCKINLEY, D. J. (1975). Nuclear polyhedrosis viruses in the control of some lepidopterous pests of tropical agriculture. Current work and thoughts on strategy. *Meded. Fac. Landb-Wet.RijksUniv.Gent,* 40: 261–265.

MORGAN, E. D., WOODBRIDGE, A. P. and ELLIS, P. E. (1975). Isolation of moulting hormones from the desert locust, *Schistocerca gregaria* (Forskal). *Acrida,* 4: 69–81. [With French summary.]

MORGAN, E. D., WOODBRIDGE, A. P. and ELLIS, P. E. (1975). Studies on the moulting hormones of the desert locust, *Schistocerca gregaria. J. Insect Physiol.,* 21: 979–993.

OVERSEAS SPRAYING MACHINERY CENTRE and CENTRE FOR OVERSEAS PEST RESEARCH (1975). Pesticide application equipment. *PANS,* 21: 436–449.

PATTERSON, J. [A.] & GOODMAN, L. J., 1975. Componental analysis of the ocellar electroretinogram of the locust, *Schistocerca gregaria. J. Insect Physiol.,* 21: 287–298.

QUINN, J. G., JOHNSTONE, D. R. and HUNTINGTON, K. A. (1975). Research and development of high and ultra-low volume sprays to control tomato leaf diseases at Samaru, Nigeria. *PANS,* 21: 388–394.

RILEY, J. R. (1975). Collective orientation in night-flying insects. *Nature, Lond.,* 253: 113–114.

RITCHIE, J. M. (1975). A revision of the grass-living genus *Podothrips* (Thysanoptera: Phlaeothripidae). *J. Ent., Lond.,* (B), 43 (1974): 261–282.

ROBERTSON, I. A. D. (1975). Entomology. *Cotton Res. Rep., Kenya,* 1973–74 (Final Report): 53–60.

ROOME, R. E. (1975). Field trials with a nuclear polyhedrosis virus and *Bacillus thuringiensis* against larvae of *Heliothis armigera* (Hb.) (Lepidoptera, Noctuidae) on sorghum and cotton in Botswana. *Bull.Ent. Res.,* 65: 507–514.

ROOME, R. E., (1975). Activity of adult *Heliothis armigera* (Hb.) (Lepiodoptera, Noctuidae) with reference to the flowering of sorghum and maize in Botswana. *Bull. Ent. Res.,* 65: 523–530.

ROOME, R. E. (1975). The control of *Heliothis* on subsistence crops in Botswana. *Meded. Fac. Landb Wet. Rijks Univ. Gent,* 40: 267–282.

SANDS, W. A. and LAMB, R. W. (1975). The systematic position of *Kaudernitermes* gen.n. (Isoptera: Termitidae, Nasutitermitinae) and its relevance to host relationships of termitophilous beetles. *J. Ent., Lond.,* (B), 44: 189–200.

WALKER, P. T. (?1973). The assessment and importance of losses caused by pests. pp.45–54. *In: Panel on pests and diseases of wheat: held at Tehran University, College of Agriculture, Karaj, Iran, February 5th–7th, 1973.* [Ankara, Turkey], Central Treaty Organization.

WALOFF, Z. and GREEN, S. M. (1975). Regularities in duration of regional desert locust plagues. *Nature, Lond.,* 256: 484–485.

WARD, P. and KENDALL, M. D. (1975). Morphological changes in the thymus of young and adult red-billed queleas *Quelea quelea* (Aves). *Phil. Trans. R. Soc. Lond.,* (B), 273: **55–64**.

WOOD, T. G. (1975). The effects of clearing and grazing on the termite fauna (Isoptera) of tropical savannas and woodlands. pp. 409–418. *In:* Vanek, J. (ed.), Progress in Soil Zoology. *Proceedings of the 5th International Colloquium on Soil Zoology, Prague, September 17–22, 1973.* The Hague, Junk.

REPORTS

COPR Report

Report of the Centre for Overseas Pest Research. January–December 1974. *Rep. Cent. Overseas Pest Res.,* 1974: vii+152pp. Price £2.30.

COPR Miscellaneous Reports

No. 17. Some aspects of the attraction of male moths of *Spodoptera littoralis* to females in the field. L. J. Rosenberg and P. M. Symmons, 1975. 10pp.

No. 18. Aerosol studies using an Aztec aircraft fitted with Micronair equipment for tsetse fly control in Botswana. C. W. Lee, G. G. Pope, J. A. Kendrick, G. Bowles and G. Wiggett, 1975. 9pp.

COPR Country Visit Reports (Distribution Limited)

COPR CVR/75/1. WHO Onchocerciasis Control Programme Projects: movements of *Similium damnosum.* J. I. Magor, 1975. 9pp. Mimeograph.

COPR CVR/75/2. A bird pest survey in the Yemen Arab Republic, November–December 1974. G. C. Pope and R. J. Douthwaite, 1975. 9pp. Mimeograph.

COPR CVR/75/3. A visit to Ghana with reference to certain environmental implications of the control of onchocerciasis and schistosomiasis in West Africa. J. Duncan and P. Matthiessen, 1975. 14pp. Mimeograph.

COPR CVR/75/4. A visit to The Gambia in connection with that country's request for the services of a malacologist (27 Nov—3 Dec 1974). J. Duncan and P. Matthiessen, 1975. 9pp. Mimeograph.

COPR CVR/75/5. Survey and control of *Patanga succincta* (The Bombay Locust) in Thailand, January—April 1975. J. Tunstall, 1975. 10pp. Mimeograph.

COPR CVR/75/6. Visit to the Sultanate of Oman to investigate the possibility of strengthening crop protection services, June 1975. H. S. Hopf, 1975. 27pp. Mimeograph.

COPR CVR/75/7. A visit to Mauritius in connection with the Natal fruit fly. E. J. Luard, 1975. 16pp. Mimeograph.

DLCOEA Technical Report

No. 63. Radiometric estimation of blood cholinesterase levels in domestic animals. R. D. MacCuaig, 1975. 8pp. Mimeograph.

FAO Reports

FAO AGPP:MISC/17. Insecticide index — 1974 (a supplement to the 1966 edition giving full details of new insecticides of possible use for locust control). R. D. MacCuaig, 1975. iv+36pp.

FAO NWA/DL/SS/1. Report of the special survey of northern Libya November 1974. G. B. Popov, 1975. [2]+29pp.

FAO UNDP/DL/TC/2. Training course on crop pest control with special reference to Desert Locust control and research. Nairobi, Kenya, 17 June — 18 July 1974. Rome, Food & Agriculture Organization of the United Nations. [3]+376pp.

The above report includes the following papers by COPR staff:

ASHALL, C. Organization, logistics, administration and evaluation of large scale pest control campaigns. pp. 271—276

BETTS, E. Locusts — past and present: biogeographical basis of warning services and control organization. pp. 277—287

JAGO, N. D. Orders and families of major insect pests of economic importance and use of keys for identification of major insect groups. pp. 24—57

MACCUAIG, R. D. Selection of insecticides for insect and control of the desert locust. pp. 145—180

POPOV, G.[B.] Desert locust survey techniques and estimation of population. pp. 114—125

RAINEY, R. C. Airborne insects and the atmospheric environment. pp. 65—81

FAO UNDP/DL/TC/4. Training course on crop pest control with special reference to desert locust control and research. Tehran, Iran 21 September — 21 October 1974. Rome, Food & Agriculture Organization of the United Nations. iv+403pp.

The above report includes the following papers by COPR staff:

MACCUAIG, R. D. Classification of insecticides and their use against various crop pests. pp. 123—131

MACCUAIG, R. D. Storage of insecticides; environmental pollution. pp. 132—139

MACCUAIG, R. D. Selection of insecticides for insect control. pp. 140—146

MACCUAIG, R. D. Control of the desert locust. pp. 147—161

FAO OSRO Reports

An interim report on the current grasshopper activity for the month of September 1975 in the Niger Basin Between Gao, Mali and Malanville, Dahomey. M. R. K. Lambert, 1975. 9+[7]pp. Mimeograph.

Report on the helicopter survey of the Niger valley from Gao to Gaya during 13 to 23 October, 1975. G. B. Popov, 1975. 20pp. Mimeograph.

An interim report on the current grasshopper activity for the month of November 1975 in the Niger Basin between Gao, Mali and Malanville, Dahomey. M. R. K. Lambert and T. O. Yonli, 1975. [25] pp. Xeroxed typescript.

A report on the grasshopper activity north of Gao, Mali during the 1975 wet season: results of surveys by members of the OSRO Crop Protection Operation Team into the Vallee du Tilemsi and west down the river Niger valley. M. R. K. Lambert and Secou Coulibaly, 1975. [19]+5+32pp. Xeroxed typescript.

ODM Reports

ODM & Government of Botswana Report. The problem of damage to sorghum by doves in Botswana, Report 1972–1974. (ODM Research Scheme R. 2664). 1975. 13pp. Mimeograph.

ODM & IITA Nigeria Report. Pesticides residues research project, 2nd interim report, 1974–75. (ODM Research Scheme R 2730). 1975. 18+6pp. Mimeograph.

ODM & University of Ibadan Report, 1975. Control of *Zonocerus variegatus* L. in Nigeria, 2nd interim report, 1973–74. (ODM Research Scheme R 2727.) 1975. [2]+48+[9]pp. Mimeograph.

WHO Reports

Annual confidential report to the World Health Organisation. Molluscicide Group of the Centre for Overseas Pest Research . . . . , 1 October 1975. 15pp. Typescript.

WHO/COPR joint studies on the movements of *Simulium damnosum* 1975. Biogeographical studies: three monthly reports, I March–May 1975; II June–August; III September–November. J. I. Magor and L. J. Rosenberg, 1975. 4; 3; 2pp. Typescript.

Windborne movement of *Simulium damnosum*. Final Report of the WHO-COPR studies in 1975. J. I. Magor, L. J. Rosenberg and D. E. Pedgley, 1975. (COPR 39/3/3). 12+[13]pp. Mimeograph.

WHO/VBC/75.510. Toxicity of insecticides to tsetse flies. A. B. Hadaway and C. R. Turner, 1975. 8pp.

MISCELLANEOUS

The Centre for Overseas Pest Research: organisation and projects. 1st revn. London, Centre for Overseas Pest Research, 1975. 12pp. Gratis.

The Centre for Overseas Pest Research: organisation and projects. 2nd revn. London, Centre for Overseas Pest Research, 1975. 12pp. Gratis.

Open days 1975: guide to exhibits. London, Centre for Overseas Pest Research, 1975. 28pp. Gratis.

FOGDEN, M. [P.L.] & FOGDEN P. (1974). *Animals and their colours. Camouflage, warning colouration, courtship and territorial display, mimicry.* London, Peter Lowe. 172pp. Price £3.75.

HODSON, M. J. and WALKER, P. T. (1975). *Stalkborer in the main maize-growing areas of Kenya.* London, Centre for Overseas Pest Research (Ministry of Overseas Development)*. [ii] +i+7pp.
*Produced by COPR for the Ministry of Agriculture, Kenya.

THESIS

LAMBERT, M. R. K. (1975). [Unpubl.]. *The flight behaviour of locusts: field and laboratory observations on the Australian plague locust,* Chortoicetes terminifera *(Walker), and the desert locust,* Schistocerca gregaria *(Forskål).* Ph.D. thesis, University of London. xii+330pp. Typescript.

3 OVERSEAS VISITS BY COPR STAFF

JANUARY

Mr J. Tunstall	Thailand	*Patanga* surveys: staff training, survey and control methods. 4 months
Mr C. Ashall	Kenya Ethiopia	To attend armyworm workshop sponsored by ICIPE. COPR liaison with EAFFRO, DLCOEA. 1 week
Dr R. C. Rainey	Kenya	Co-Chairman of ICIPE/EAAFRO/COPR joint study workshop on armyworm; paper presented. 1 week
Mr P. T. Walker	Antigua St. Lucia Martinique Guadeloupe Puerto Rico	Survey of sweet potato weevil damage and control. Contact with French research workers. 3 weeks
Mr A. L. Davies	Nigeria	To study chemical control of *Zonocerus variegatus;* bioassay trials on termites at Mokwa. 3 months
Dr D. G. Campion	Kenya Tanzania	To attend the armyworm study workshop organised by ICIPE; to initiate trials on sex pheromone control of East African armyworm. 3 weeks
Dr W. A. Sands	Kenya	Supervision of ecological research on grass-feeding termites at ICIPE. 2 weeks
Mr R. A. Steedman	Thailand	To provide technical assistance on control of Bombay Locust, *Patanga*. 8 months
Mr D. R. Johnstone Mrs K. A. Johnstone	Swaziland South Africa	To test aerial spray equipment. To attend CRC workshop 'Ultra-low volume spraying for cotton pest control'. To liaise with National Bureau of Standards, Pretoria. 1½ months
Dr R. C. Rainey	Canada	Co-ordination of field research programme at Maritime Forest Research Centre. 12 days

FEBRUARY

Dr R. C. Rainey	Sudan	To present paper at seminar on strategy for cotton pest control in the Sudan Gezira region. 1 week
Dr R. F. Chapman Mr W. W. Page Mr E. G. Harris Dr P. Hunter-Jones Dr A. R. McCaffery Dr E. A. Bernays	Nigeria Ghana	To participate in *Zonocerus* scheme. 5 weeks
Mr C. W. Lee	Switzerland	To liaise with FAO/WHO on insecticide control of tsetse fly. 3 days
Mr G. G. Pope	Botswana	To assist COPR Dove Project. 5 weeks

MARCH

Dr L. J. Rosenberg Dr J. I. Magor	Upper Volta Ghana Togo Ivory Coast	Biogeographical studies on *Simulium damnosum*, a WHO/COPR project; collection of meteorological data. 9 months
Mr D. J. McKinley Dr D. G. Campion	Belgium	To attend 27th International Symposium on Crop Protection at Rijksuniversiteit Ghent. 2 days
Mr C. F. Hemming	Mali	To attend seminar 'The Evaluation and Mapping of Tropical African Rangelands' and visit OICMA for discussions. 2 weeks
Mr T. Jones	Senegal	To attend the conference of the Association for the Advancement of Agricultural Sciences in Africa. 1 week

APRIL

Mr T. J. Perfect	Rome	To attend FAO/UNEP Expert Consultation on Impact Monitoring of Residues from agricultural pesticides in developing countries. 5 days
Dr D. G. Campion	Egypt Greece	**Consultant to FAO Near East ad hoc Joint Consultation on pest and disease appraisal and control by management systems;** discussions on sex pheromone control of *Spodoptera littoralis*. 3 weeks

Mr D. E. Pedgley	Upper Volta	Examination of weather systems at times of distribution changes of *Simulium damnosum*. 6 weeks
Mr C. Ashall	Italy	Member of UK delegation to ad hoc Government Consultation on Pesticides in Agriculture and Public Health (FAO). 5 days
Mr J. E. H. Grose Mr E. G. Harris	Netherlands	To attend International Symposium on the Evaluation of Biological Activities, Wageningen. 3 days
Miss A. G. Cook Mr T. J. Perfect	Nigeria	To co-operate with IITA, Ibadan investigating effect of DDT on the soil ecosystem. 8½ months
Mr B. W. Bettany Mr P. T. Walker Mr M. R. Hodgson	Kenya	Extension exercise on maize stalk borer. Lectures and film. 8 weeks
Mrs A. Steedman	Thailand India Pakistan	To collect data for the Desert Locust Forecasting Manual Project. 4½ months
Miss P. Wortley	Mexico	To attend the 4th Inter-American and the World Congress of Agricultural Librarians and Documentalists. 12 days
Mr R. J. Douthwaite	Thailand	To evaluate field studies of locust nymphs and adults. 3½ months
Mr C. Longhurst	Nigeria	To collect live specimens for research purposes. 2 months
Dr P. T. Haskell	Rome	UK delegate to FAO Pesticide Conference. 5 days
Mr R. L. Moore	Nigeria	Field work with pesticide residues team in Ibadan; effect of DDT on soil fertility. 2 months
Mr T. Jones	India-Pakistan	To attend the International Union of Forest Research Organisations/FAO Conference; liase with ICRISAT, CIBC. 2 weeks
Mr R. M. C. Williams	Berlin (West)	To attend symposium 'Organismen und Holz', sponsored by Bundesanstalt fur Materialprufung, Dahlem, W. Berlin. 2 days

MAY

Dr T. G. Wood	Nigeria	Supervise arrangements for growing season of experimental plots for Termite Research Project at Mokwa. 3 days
Dr D. G. Campion	Iran	Member of OLLB Working Group on integrated control of cotton pests. 9 days
Dr D. G. Campion	Belgium	To attend International Symposium on Crop Protection. 2 days
Dr W. A. Sands	Kenya	To attend Annual Research Conference of ICIPE, and supervise termite ecological research. 3 weeks.
Mr G. Popov Dr M. R. K. Lambert Miss P. McAleer	West Africa	To serve as co-ordinater with OSRO Emergency Crop Protection Operations, to attend FAO/UNDP training course on Desert Locust and Pest Control. 8 months
Dr R. E. Roome	Belgium	To attend XXVII International Symposium on Crop Protection. 2 days
Dr P. M. Symmons	Gabarone	To attend the fifth Ordinary Session of Governing Council of IRLCOCSA. 4 days
Dr A. B. Hadaway	Italy	To attend the FAO Expert Consultation on research on tick-borne diseases and their vectors. 5 days
Dr P. T. Haskell Dr P. Rosen	France	Discussions with OECD. 1 day
Dr P. T. Haskell Dr B. Steele	Belgium	To attend the International Plant Protection Symposium as chairman of Special Session. 4 days

JUNE

Mr J. Tunstall	Zambia Tanzania	Red Locust Survey for IRLCOCSA. 3 months
Dr R. C. Rainey	Canada	To assess and co-ordinate research on spruce budworm control particularly by aerial spraying. 1 month
Mr J. Roffey	Italy	Briefing by FAO Remote Sensing Unit; project to detect desert locust breeding areas by satellite. 3 days

Dr N. D. Jago	Cameroons	To attend OICMA Council Meeting. 2½ weeks
Mr M. Hodson	Upper Volta Ghana	To investigate problems on the WHO/COPR River Blindness Project. 6 weeks
Miss J. M. Child	France	To hold discussions with Association d'Acridologie about *Acrida*. 3 days

JULY

Mr R. M. C. Williams	Ghana	To the Building and Road Research Institute to advise on current and projected termite materials — testing work. 2 weeks
Mr J. Roffey	USA	Visit to NASA, National Oceanic and Atmospheric Administration to discuss detection of desert locust breeding areas by satellite. 5 days
Dr D. G. Campion	Morocco Crete Pakistan	To attend the FAO/UNEP Consultation on pest management systems for the control of cotton pests; to initiate projects on pheromone control of Egyptian cotton pest, *Spodoptera littoralis*. 3½ months
Mr L. McVeigh	Crete	To participate in project for pheromone control of Egyptian cotton pest *Spodoptera littoralis*. 4 months
Dr J. Murlis	Crete	To study behaviour of the cotton pest *Spodoptera littoralis* for pheromone control. 1 month
Dr P. Hunter-Jones Dr P. M. Symmons Mr J. Tunstall	Tanzania	Field study on insecticide control of Red Locust. 3 weeks
Mr C. W. Lee	Botswana	To evaluate insecticide control of tsetse fly by aircraft spraying. 1 month
Mr W. T. King	Gambia	To implement ULV pesticide application field trials; to lecture and demonstrate. 4½ months

AUGUST

| Dr R. C. Rainey
Dr D. G. Campion | USSR | Present paper at 8th International Plant Protection Congress in Moscow as member of Presidium of Congress.
12 days |

Dr T. G. Wood	Nigeria	Supervise termite research project at Mokwa. 2 weeks
Miss S. C. A. Cook	USA Belize Honduras Costa Rica Colombia Equador	To attend the World Soybean Research Conference, collect information for revision of PANS Manual 'Pest Control in Bananas'. 1 month
Mr A. Davies	Nigeria	Bioassay trials on termites, training students. 3 weeks
Dr J. Duncan	Sudan Swaziland	To attend discussions with Commonwealth Development Corporation on bilharzia control particularly by aerial spraying technique. 12 days
Miss E. Luard	Mauritius S. Africa	To assess the Natal fruit fly problem.
Mr C. Foss	Ethiopia	To attend the Review Commission of the Desert Locust Control Organisation — Eastern Africa as a consultant on conditions of service. 7 days

SEPTEMBER

Mr J. Roffey	Italy	To attend the 19th Seminar of the Desert Locust Control Committee (FAO) and technical discussions on locust control by satellite. 10 days
Miss A. M. Ward	Belgium	To attend Semaine d'Etude Agriculture et Hygiene des Plantes Gembloux. 6 days
Dr R. J. Cooter Mr P. S. Baker	Germany	To attend 1st European Neuro-sciences Meeting, Munich and visit the Zoology Dept., University of Konstanz. 10 days
Dr W. A. Sands	Sri Lanka	To discuss Kalotermitidae damage to tea. 2½ weeks
Dr N. D. Jago	Ethiopia	Liason with FAO team controlling grasshopper pests. 3 months
Dr P. Rosen	France	To review progress of OECD Report on Pest and Vector Control. 3 days
Dr P. Hunter-Jones	Nigeria	Insecticide trials to control *Zonocerus variegatus*. 5 weeks

Dr D. R. Reynolds	Upper Volta	A familiarisation visit to *Simulium damnosum*/onchocerciasis project. 4 weeks
Mr R. A. Steedman	Malaysia	To discuss status of *Valanga nigricornis*. 1 week
Mr J. Beesley	Antigua	To assess research results of dove research scheme. 4 weeks
Dr P. T. Haskell	Kenya	Executive discussions with EADD, UNEP, ICIPE, EAAFRO. 5 days
Dr P. Rosen	Rome	To review progress on OECD Report on Pest and Vector Control at meeting with FAO, World Food Council, OSRO. 4 days

OCTOBER

Dr R. Reynolds Mr A. D. Smith Dr J. Murlis	Mali	To conduct Radar observations of the migratory locust in association with OICMA. 10 weeks
Dr W. Reed	Pakistan	Consultant to Director at FAO/UNEP Meeting. 2 weeks
Dr R. F. Chapman	Netherlands	To liaise with Agricultural University, Wageningen on subjects of mutual interest. 3 days
Dr P. Rosen	USA Canada	To assess OECD Report on research needs in pest and vector control with World Bank, UNDP. 9 days
Dr E. A. Bernays	Netherlands	To liaise with Agricultural University Wageningen on topics of mutual interest. 3 days
Dr P. T. Haskell	Pakistan	To attend the UNEP/FAO Cotton Pest Conference. 11 days
Mr D. R. Johnstone	Thailand	To liaise on cotton pest control; field studies on ULV application of pesticides. 2 weeks
Dr P. Rosen	France	To attend meeting of Steering Group of OECD to discusss progress of the report 'Pest and Vector Control in Developing Countries.' 4 days
Dr P. Ellis	Crete	To assess sampling methods of *Spodoptera littoralis*. 2 weeks

Mr D. J. McKinley	Crete	Field work on the nuclear polyhedrosis virus of *Spodoptera littoralis*. 1 month
Dr J. Duncan	Egypt	To attend the International Conference on Schistosomiasis. 9 days
Mr A. Mitchell	Nigeria	Field trials to study insecticide control *Zonocerus variegatus*. 1 month
Mr C. Ashall	Ethiopia	Technical Committee of DLCOEA, liaison with Ethiopian Plant Protection Dept., discussions on armyworm control. 1 week
Mr C. Ashall	Italy	Paper presented at XIXth Session of Desert Locust Control Committee (FAO).
Dr R. C. Rainey	Sudan	Participation in Sudan Gezira Research Project on air-borne pests. 1 month

NOVEMBER

Dr H. S. Hopf	Brazil Antigua Paraguay Costa Rica Barbados St. Lucia	Liaison with West Indies crop protection projects and advisory missions. 4 weeks
Mr C. Ashall Mr P. T. Walker	Bangladesh	To advise Bangladesh Government on pest control development. 4 weeks
Mr J. E. H. Grose	Kenya Mauritius S. Africa	Discussions on biting fly control, consultant on *Stomoxys nigra* control. 2 weeks
Dr R. F. Chapman Dr G. A. Mitchell Dr E. A. Bernays Mr W. W. Page	Nigeria	To study population, causes of mortality of *Zonocerus*. To attend the 10th Anniversary Celebrations of the Nigerian Entomological Society. 9 weeks
Mr J. P. Tunstall	Pakistan	To initiate a UK aid project to control cotton pest. 3 weeks
Mr G. G. Pope Mr C. W. Lee	Botswana	To conduct tsetse spraying; trials with hovercraft. 1 month
Mr T. Jones	Thailand Malaysia Indonesia	Liase with SEADD, to discuss involvement in plant protection programme. 1 month

Dr P. Rosen	Holland	To hold discussions with scientific institutes on the OECD Steering Group Report on Pest Control in Small Farmer Food Crops. 1 week

DECEMBER

Dr T. Wood	Nigeria	Supervise Termite Research Project at Mokwa. 10 days
Mr C. Foss	Nigeria	To discuss with International Institute of Tropical Agriculture, Ibadan the secondment of COPR staff.
Dr P. T. Haskell	France	To attend OCP Meeting. 2 days

4 CONFERENCES AND MEETINGS ATTENDED BY COPR STAFF

JANUARY

International Centre for Insect Physiology and Ecology (ICIPE) Workshop on Armyworm in Nairobi

Mr D. J. McKinley contributed a paper on current research on nuclear polyhedrosis virus of *Spodoptera*

FEBRUARY

University College, London Zoological Society Meeting in London

Dr M. R. K. Lambert lectured on "The flight activity of locusts"

International Institute for Tropical Agriculture In-house Review meeting at Ibadan, Nigeria

Dr R. F. Chapman gave a talk on "Work of the COPR Pesticide Ecology Group"

MARCH

British National Farmers' Union British Growers Look Ahead Conference and Exhibition at Harrogate, England

Miss J. M. Child, Miss A. Ward, Miss S. C. A. Cook and Mr S. Mercer attended and manned the COPR exhibit of publications

Cotton Research Corporation, London Workshop "ULV spraying for cotton pest control" at Big Bend, Swaziland

Mr D. R. Johnstone and Mrs K. A. Johnstone attended and advised on production of the report

Imperial College, London Seminar on "The flight activity of locusts" at Silwood Park, Ascot, England

Dr M. R. K. Lambert attended

APRIL

British Ecological Society Symposium on "The role of terrestrial and aquatic organisms on the decomposition process" at Coleraine, Ireland

Dr T. G. Wood gave a paper entitled "The role of termites in the decomposition process"

University of Ghana, Department of Zoology Research Seminar at Legon

Dr R. F. Chapman gave a lecture on "The biology and control of *Zonocerus variegatus*"

Nigerian Field Society Meeting at the University of Ibadan

Dr R. F. Chapman gave a talk on "The biology and control of the variegated grasshopper"

International Symposium on the Evaluation of Biological Activity at Wageningen, Netherlands

Mr E. G. Harris attended Reported in *PANS* 21(3): 335—339

East Malling Research Station Members' Day on apple mildew at Maidstone, England	Miss S. C. A. Cook attended Reported in *PANS* 21(3): 315—6
Association of Animal Behaviour Seminar on Pheromones Workshop on Population Modelling in London	Dr J. Murlis attended
Royal College of Art Symposium "Design for Need"	Dr J. Murlis read a paper on the role of designers in disaster relief.

MAY

International Centre for Insect Physiology and Ecology Annual Research Conference in Nairobi	Dr W. A. Sands gave an introductory paper on the work of the termite ecology group
Food and Agriculture Organization of the UN Expert Consultation on "Research on Tick-borne Diseases and their Vectors" in Rome	Dr A. B. Hadaway attended
Joint FAO/Industry Task Force First Session on "Ticks and Tick-borne Disease Control" in Rome	Dr A. B. Hadaway attended
University of Ghent, Faculty of Agriculture 27th International Symposium on Crop Protection at Ghent, Belgium	Dr P. T. Haskell organised and chaired the special session on 'Tropical Crop Protection' The following papers were presented: "The control of *Heliothis* on subsistence crops in Botswana" — Dr R. E. Roome "Laboratory evaluation of the activity of synthetic pyrethroids at ULV against the cotton leafworm, *Spodoptera littoralis*" — Dr M. Ford, Dr. C. Reay and Mr W. S. Watts

JUNE

Royal Agricultural Society of England The Royal Show at Kenilworth, England	Miss A. Ward, Mr S. Mercer and Miss S. C. A. Cook attended Reported in *PANS* 21(3): 431—435
Kent Beekeepers' Association Meeting at Tonbridge, England	Dr R. F. Chapman gave a talk entitled "How insects taste and smell"
Association of Applied Biologists Meeting on sugarbeet pests at Bury St. Edmunds, England	Mr P. T. Walker attended

JULY

Commonwealth Agricultural Bureaux
First Commonwealth Helminthological
Meeting
at St. Albans, England

Miss A. Ward and Miss S. C. A. Cook attended
Reported in *PANS* 21(4): 422—425

ARC Weed Research Organization Meeting
on "Herbicide application in very low
volume"
at Oxford, England

Mr D. R. Johnstone and Mrs K. A. Johnstone
attended and gave a paper on theoretical and
practical considerations of spraying at ULV
rates

Commonwealth Agricultural Bureaux
Ninth Commonwealth Conference on
Plant Pathology
at St. Albans, England

Miss S. C. A. Cook attended

AUGUST

University of Illinois
World Soybean Research Conference
at Urbana-Champaign, USA

Miss S. C. A. Cook attended
Reported in *PANS* 22(1): 116—123

SEPTEMBER

Association of Applied Biologists Meeting
on "Pests and Climate"
at Bath, England

Mr P. T. Walker and Mr M. J. Hodson read a
paper on "The effect of climate on the distribution and life-cycle of the maize stem-borer,
Busseola fusca, in East Africa"

Royal Entomological Society
Symposium on Insect Development
at Imperial College, London

Miss E. Betts and Dr R. F. Chapman attended

Fifth International Agricultural Aviation
Conference
Stoneleigh, Kenilworth, England

Dr A. B. Hadaway and Mr W. S. Watts attended.
Dr Hadaway and Mr D. R. Johnstone were
organizers and Chairmen of the session on
"Aviation in Public and Animal Health"
Mr C. W. Lee presented a paper by J. Parker,
D. Baldrey and C. W. Lee on "The use of aircraft in the WHO Onchocerciasis Programme"

Food and Agricultural Organization of the
UN 19th Session of the Desert Locust
Control Committee
in Rome

Mr C. Ashall and Mr J. Roffey attended as UK
delegates

FAO of the UN
Technical Consultation on the Project Plan
for the use of satellite techniques in
Desert Locust survey
in Rome

Mr J. Roffey attended

OCTOBER

Royal Society
Conference on "Quality and Economics of
Scientific Journals"
in London

Mr C. F. Hemming and Mr S. Mercer attended

Society of Chemical Industry
Symposium on "Droplets in air, Part I,
The generation and behaviour of airborne
dispersions."
in London

Mr D. R. Johnstone and Mrs K. A. Johnstone
attended

NOVEMBER

ELSE/EDITERRA
Workshop on 'Techniques of Typesetting
and Printing'
in London

Mr C. F. Hemming, Miss A. Ward, Miss S. C. A.
Cook and Mr S. Mercer attended.

British Crop Protection Council
8th Insecticide and Fungicide Conference
at Brighton, England

Mr P. T. Walker, Miss A. Ward and Miss S. C. A.
Cook attended.
Dr P. J. Jones delivered a paper entitled "The
significance of bird migration in relation to bird
pest control strategy"
Mr D. R. and Mrs K. A. Johnstone presented a
paper entitled "Physical assessment of very low
volume fungicide spray on tomatoes in northern
Nigeria"
Report in *PANS* 22(2): 301–304

DECEMBER

Entomological Society of Nigeria
10th Anniversary Celebrations
at University of Ibadan

Dr R. F. Chapman was invited to speak on
"Entomology in Society"

The Royal Smithfield Show
in London

Miss A. Ward and Miss S. C. A. Cook attended

Society of Chemical Industry,
Pesticides Group
Meeting on "Control of Disease Vectors
in the Public Health Field"
in London

Dr P. Mathiessen attended
Dr A. B. Hadaway presented a paper on
"The search for new insecticides for tsetse
control"
Dr J. Duncan presented a paper on
"Problems in the control of schistosomiasis"

5 COMMITTEES, COPR REPRESENTATION IN 1975

The Director represented ODM on the following:

British Council, Science Advisory Committee
British Crop Protection Council
Royal Society, Biological Control Sub-committee
Advisory Board to the Research Councils, Working Party on Taxonomy

he represented COPR on the following:

ODM Natural Resources Advisers Co-ordinating Committee
ODM Management Committee for COPR

and was a member of the following:

WHO/IBRD Onchocerciasis Project, Scientific and Technical Advisory Committee
WHO Committee on Insecticides (Pest Ecology)
Board of Governors ICIPE
Royal Society UK National Committee for ICIPE
Council of the Association d'Acridologie
ODM Trypanosomiasis Panel
ODM NRACC Sub-committee on Overseas Pesticide Application
(Dr Haskell Chairman, Mr C. W. Lee Secretary)
ODM NRACC Sub-committee on Environment
Advisory Committee David Owen Centre for Population Growth Studies, University College, Cardiff

Other staff members represented COPR as follows:

ODM Scientific Units Information Retrieval Computerisation Working Party.
(Mr C. F. Hemming, Miss P. J. Wortley)
ODM Superannuation Scheme Advisory Board (Dr W. A. Sands)
MAFF Pesticides Analysis Advisory Committee, Petroleum Oils Sub-committee
(Mr F. Barlow)
MAFF Pesticides Analysis Advisory Committee, Formulation Panel
(Mr F. Barlow)
MAFF Pesticides Analysis Advisory Committee, Formulation Panel (Granule Group)
(Mr P. T. Walker)
British Standards Institution Technical Committee PCC/1 on Common Names for Pesticides
(Mr F. Barlow)
British Standards Institute Committee WPC/2, Classification of Wood Preservatives
(Mr R. M. C. Williams)
BSI Committee WPC/10 Wood Preservation Tests (Mr R. M. C. Williams)
Fifth International Agricultural Aviation Congress Programme Committee
(Dr A. B. Hadaway)
Fifth International Agricultural Aviation Congress Organising Committee
(Mr D. R. Johnstone, Dr R. C. Rainey)
FAO Desert Locust Control Committee (Mr C. Ashall, Mr J. Roffey)
FAO/IAEA Division Technical Advisory Group to Insect Pest Control Section
(Dr D. G. Campion)
WHO Committee on Insecticides (Application & Dispersal of Pesticides) (Mr C. W. Lee)
WHO Representatives of Collaborating Laboratories (Dr A. B. Hadaway)
WMO Working Group on Meteorological Aspects of Aerobiology (Mr D. E. Pedgley)

FAO Working Group on Crop Losses caused by Pest, Diseases and Weeds
  (Mr P. T. Walker)
Royal Entomological Society of London Library Committee (Dr R. F. Chapman)
    Symposium Committee (Dr R. C. Rainey)
    Finance and House Committee (Dr W. A. Sands)
Council of Association D'Acridologie (Mr C. F. Hemming)
Institute of Biology, Committee on Modification of Bye-laws (Dr W. A. Sands)
Institute of Biology, Examining Board (Dr R. F. Chapman)
Institute of Biology, Careers Committee (Mr A. T. Thompson)
ODM NRACC Sub-committee on Overseas Pesticide Application (Dr A. B. Hadaway)
WHO Scientific Advisory Panel concerned with Onchocerciasis Control in the Volta River
  Basin (Mr C. W. Lee)
ICIPE Policy Advisory Committee (Dr W. A. Sands)
Royal Meteorological Society. Honorary Secretary (Mr D. Pedgley)
Brighton Polytechnic, Advisory Board for Biology (Dr R. F. Chapman)

6 COPR CONSULTANTS

The Centre received advice from the following Consultants during 1975:
Dr N. Waloff
  Imperial College, London
Prof J. R. Busvine
  London School of Hygiene and Tropical Medicine
Dr T. Swain
  ARC Unit of Plant Biochemistry, Kew
Prof T. R. E. Southwood
  Imperial College
Dr C. Edwards
  Rothamsted Experimental Station
Dr J. M. Baker
  Forest Products Laboratory, Princes Risborough
Dr I. J. Graham-Bryce
  Rothamsted Experimental Station
Mr G. M. Jolly
  ARC Unit of Statistics, Edinburgh
Dr W. R. Woof
  University of Salford
Mr K. J. R. MacLennan
Dr J. Ford
Dr D. L. Gunn

7 VISITORS TO COPR

The following is a list of some of the more important visitors to COPR in 1975:

International and regional

Association d'Acridologie (Paris) Dr F. O. Albrecht (President)
OICMA (Mali)   M. Gana Diagne (Director General)
WHO (Geneva)   Dr A. R. Stiles
               Dr N. Gratz
               Mr J. D. Parker
ICIPE (Nairobi) Dr G. Prestwick

Angola

Museo do Dundo   Dr A. de Barros Machado

Australia

CSIRO   Mr C. R. Miller (Division of Forests)
        Mr R. Trayford

Botswana

Department of Veterinary Services and Tsetse Control   Mr G. Bowles
                                                       Dr J. Davis

Ministry of Agriculture   Mr E. H. Watt
                          Miss K. E. Flattery

France

Roussel-Uclaf   Dr B. Leraillez

Germany

Bundesanstalt für Materialprüfung   Prof. G. Becker
Hoechst Ltd.   Dr E. F. Schultze
               Mr N. Alsop

Ghana

Legon University, Accra   Mrs L. McGowan

India

University of Delhi   Prof. K. Saxena

Israel

Volcani Institute   Dr P. Vermes
                    Dr M. Sternlicht
Hebrew University   Dr P. Pener

Japan

Sumitomo Chemical Co. Ltd.   Dr J. Hattori

Kenya

Pyrethrum Marketing Bureau, Nakuru   Mr G. D. Glynne-Jones
                                      Mr R. Winney
Hoechst (East Africa) Ltd.   Mr D Grieve
Kenyatta College, University of Nairobi   Dr D. Pomeroy

Malawi

Tea Research Institute   Dr Pritam S. Rattan

Malaysia

Forest Research Institute, Kepong   Mr Tho You Pong

Mexico

National University, Mexico City   Dr L. M. Pinzóu Picanseño
                                    Dr V. Perez Morales

Nigeria

Institute for Agricultural Research, Zaria   Mr J. G. Quinn
                                              Dr Ochappa Onazi

Senegal

FAO Horticultural Scheme, Dakar   Mr L. Collingwood

Swaziland

University of Botswana, Lesotho   Dr G. Rohrmann

Switzerland

Ciba-Geigy Ltd.   Dr A. Renfer

USA

Utah State University   Dr T. Hsaio
University of California   Prof E. B. Edney
FMC International Ltd.   Mr D. M. Boyd
Hudson Manufacturing Co.   Mr R. Treichler
Smithsonian Institute, Washington   Dr H Tyson-Roberts
USDA, Gulfport   Dr F. L. Carter

Great Britain

Weed Research Organization   Mr C. R. Merritt
                              Mr W. A. Taylor
Information Research Ltd.   Dr K. Quarmby
Tsetse Research Laboratory, Langford   Dr A. N. Jordan (Director)
Royal Botanic Gardens   Dr T. Swain
                         Dr G. Cooper
Animal Virus Research Institute   Dr R. F. Sellers
Rothamsted Experimental Station   Dr J. Bowden
Biodeterioration Centre, Birmingham   Prof T. E. Oxley
ICI Plant Protection Ltd.   Miss P. Woolner
                             Dr A. Green
                             Mr R. Dorner

Dow Chemical Co.	Mr P. R. Sparrow
	Mr R. Dutton
May and Baker Ltd.	Mr H. J. Cottrell
	Mr B. H. Betts
Fisons Ltd., Chesterford Park	Dr R. W. Lemon
	Dr C. R. Bromilow
	Mr G. Thompson
Wellcome Research Laboratories	Dr G. Dodd
	Mr M. Matthewson
	Mr G. Blackman
	Dr P. R. Chadwick
	Mr J. Wood
	Mr J. C. Wickham
	Mr J. P. Brook
Shell International	Mr M. O. Lane
Shell Research Ltd.	Dr J. C. Felton
	Mr H. H. Coutts
Hoechst Ltd.	Mr A. M. Jones
ICI Agricultural Division, Billingham	Mr S. B. Gates
	Mr R. Veitch
Muirhead and Sons, Grangemouth	Mr. D. Pringle
	Mr M. Clark
Tinsley Wire Industries	Mr A. F. Bull
Building Research Establishment, Princes Risborough	Mr J. M. Baker
Chelsea College, Univ. of London	Dr D. Payne
University of Oxford	Prof. G. C. Varley
	Dr P. Miller
Sunderland Polytechnic	Dr R. Barass
York University	Dr M. B. Usher
	Dr S. Gillett
Exeter University	Dr J. Anderson
Keele University	Dr E. D. Morgan
	Mr M. Poole
Portsmouth Polytechnic	Dr R. Reay
	Dr M. Ford
	Dr R. Greenwood
Birkbeck College, University of London	Dr W. M. Blaney
	Miss C. Winstanley
Imperial College, Department of Aeronautics	Prof P. R. Owen
	Dr A. Taylor-Russell
	Dr M. R. Graham
	Dr J. Harvey
	Dr D. Hillier

8 GLOSSARY

AAASA Association for the Advancement of Agricultural Sciences in Africa

ADAS Agricultural Development and Advisory Service, MAFF

AID Agency for International Development (US)

ALRC Anti-Locust Research Centre (UK)

ARC Agricultural Research Council (UK)

ARI Agricultural Research Institute (Cyprus)

ASECNA Agence pour la Securité de la Navigation Aerienne en Afrique et Madagascar

ASLIB Association of Special Libraries and Information Bureaux (UK)

BCPC British Crop Protection Council

BRE Building Research Establishment, DOE

BREPRL Building Research Establishment, Princes Risborough Laboratory

BRRI Building and Road Research Institute (Ghana)

CATIE Centro Agronomico deo Investigation y Ensenanza (Costa Rica)

CCA Central Computer Agency, CSD

CENTO Central Treaty Organisation (Iran)

CIAT International Centre of Tropical Agriculture (Colombia)

CIBC Commonwealth Institute for Biological Control

CIDA Canadian International Development Agency

CIE Commonwealth Institute of Entomology

CIH Commonwealth Institute of Helminthology

CMI Commonwealth Mycological Institute

CNRS Centre Nationale Recherche Scientifique (France)

CIPAC Collaborative International Pesticides Analytical Committee

CPPTI Central Plant Protection Training Institute (India)

COPR Centre for Overseas Pest Research (UK)

CSD Civil Service Department (UK)

CSIRO Commonwealth Scientific and Industrial Research Organisation (Australia)

CYMMIT International Maize and Wheat Improvement Centre (Mexico)

DLCC Desert Locust Control Committee (FAO)

DLCOEA Desert Locust Control Organization for Eastern Africa

DLIS Desert Locust Information Section, COPR

DOE Department of the Environment (UK)

EAAFRO East African Agriculture and Forestry Research Organization

EAC East African Community
ECRO European Chemoreception Research Organization
ELSE European Association of Editors of Biological Periodicals
EPPO European and Mediterranean Plant Protection Organization
ERTS Earth Resources Technology Satellite
ESC European Standards Committee
FAO Food and Agriculture Organization of the United Nations
FPRI Forest Products Research Institute (Ghana)
FPRL Forest Products Research Laboratory (UK)
GIFAP Groupement International des Associations Nationales des Fabricants de Pesticides
IAAC International Agricultural Aviation Centre (Netherlands)
IAALD International Association of Agricultural Librarians and Documentalists
IAC International Agricultural Centre (Netherlands)
IAEA International Atomic Energy Authority (Vienna)
IAR Institute for Agricultural Research (Samaru, Nigeria)
IARI Indian Agricultural Research Institute
IBRD International Bank for Reconstruction and Development
IC Imperial College of Science and Technology, University of London
ICAR Indian Council for Agricultural Research
ICIPE International Centre for Insect Physiology and Ecology (Kenya)
ICRA International Copper Research Association Inc.
ICRISAT International Crops Research Institute for the Semi-Arid Tropics (India)
IDRC International Development Research Centre (Canada)
IITA International Institute of Tropical Agriculture (Nigeria)
ILCA International Livestock Centre for Africa
INTA National Institute of Agricultural Technology (Argentina)
IOBC International Organization for Biological Control
IOBC/WPRS West Palearctic Regional Section of IOBC
IRLCOCSA International Red Locust Control Organization for Central and Southern Africa
IRRI International Rice Research Institute (Philippines)
ISO International Standards Organization
IUSSI International Union for the Study of Social Insects
LRD Land Resources Division, ODM
LSHTM London School of Hygiene and Tropical Medicine

MAFF   Ministry of Agriculture, Fisheries and Food (UK)
MAGN   Ministerio de Agricultura y Ganadia de la Nacion (Argentina)
NAL   National Agricultural Laboratories (Kenya)
NASA   National Aeronautics and Space Administration (USA)
NERC   Natural Environment Research Council (UK)
NRACC   Natural Resources Advisers Coordinating Committee of ODM
OCCGE   Organisation de Coordination et de Cooperation pour la lutte Contre les Grandes Endemies en Afrique de l'Ouest (Upper Volta)
OCLALAV Organisation Commune de Lutte Antiacridienne et de Lutte Antiaviare (Senegal)
OCP   Onchocerciasis Control Programme (West Africa)
ODM   Ministry of Overseas Development (UK)
OECD   Organization for Economic Cooperation and Development (France)
OICMA   Organisation Internationale contre le Criquet Migrateur Africaine (Mali)
ORSTOM Office de la Recherche Scientifique et Technique Outre-Mer (France)
OSMC   Overseas Spraying Machinery Centre, Imperial College (UK)
OSRO   Office of Sahelian Relief Operations
PICL   Pest Infestation Control Laboratory, MAFF
PUDOC   Publication and Documentation Centre (Netherlands)
RRE   Royal Radar Establishment (UK)
SCI   Society for Chemical Industry (UK)
SIDA   Swedish International Development Authority
SIE   Secretariat for International Ecology (Sweden)
TNO   Organisatie voor Toegepast-Naturwitenschapelijk Ondezoek (Netherlands)
TPI   Tropical Products Institute, ODM
TPRI   Tropical Pesticide Research Institute (Tanzania)
TSPC   Tropical Stored Products Centre, ODM
UN   United Nations
UNDP   United Nations Development Programme
USDA   United States Department of Agriculture
WMO   World Meteorological Organization of the United Nations
WHO   World Health Organization of the United Nations
WRO   Weed Research Organization (UK)